Structure, Evidence, and Heuristic

"Schulz addresses the value of the emerging interdisciplinary field of evolutionary economics by considering its contributions in three main areas—the structural, evidentiary, and heuristic. The result is a clear-eyed and thoughtful assessment of the merits of evolutionary economics that highlights both the contexts in which it is useful and those in which it is not. This book will be valuable for shaping the direction of the field, and to anyone who is interested in a strong framework for assessing a newly emerging interdisciplinary field."
—**Sarah F Brosnan**, *Georgia State University, USA*

This book is the first systematic treatment of the philosophy of science underlying evolutionary economics. It does not advocate an evolutionary approach towards economics, but rather assesses the epistemic value of appealing to evolutionary biology in economics more generally.

The author divides work in evolutionary economics into three distinct, albeit related, forms: a structural form, an evidential form, and a heuristic form. He then analyzes five examples of work in evolutionary economics falling under these three forms. For the structural form, he examines the parallelism between natural selection and economic decision-making, and the parallelism between natural selection and market competition. For the evidential form, he looks at the relationship between animal and human economic decision-making, and the evolutionary explanation of diversity in human economic decision-making. Finally, for the heuristic form, he focuses on the plausibility of equilibrium modeling in evolutionary ecology and economics. In this way, he shows that linking evolutionary biology and economics can make for a powerful methodological tool that can enable progress in our understanding of various economics questions.

Structure, Evidence, and Heuristic will be of interest to scholars and advanced students working in philosophy of science, philosophy of social science, evolutionary biology, and economics.

Armin W. Schulz is Associate Professor of philosophy at the University of Kansas. His research concerns the implications of evolutionary biological considerations for the social and cognitive sciences. He is the author of *Efficient Cognition: The Evolution of Representational Decision Making* (2018), and over 20 published papers.

Routledge Studies in the Philosophy of Science

For more information about this series, please visit: www.routledge.
com/Routledge-Studies-in-the-Philosophy-of-Science/book-series/POS

Structure, Evidence, and Heuristic

Evolutionary Biology, Economics, and the Philosophy of Their Relationship

Armin W. Schulz

Routledge
Taylor & Francis Group

NEW YORK AND LONDON

First published 2020
by Routledge
52 Vanderbilt Avenue, New York, NY 10017

and by Routledge
2 Park Square, Milton Park, Abingdon, Oxon OX14 4RN

Routledge is an imprint of the Taylor & Francis Group, an informa business

© 2020 Taylor & Francis

Library of Congress Cataloging-in-Publication Data
A catalog record for this book has been requested

ISBN: 978-0-367-46590-2 (hbk)
ISBN: 978-0-367-49254-0 (pbk)
ISBN: 978-1-003-03024-9 (ebk)

Typeset in Sabon
by Apex CoVantage, LLC

For Kelly, James, and Elizabeth—we are a team.

Contents

Figures

Tables

Glossary of Abbreviations

Abbreviation	Description
RCT	"Rational Choice Theory": the classic theory of economic decision-making according to which economic agents maximize their expected utilities
PT	"Prospect Theory": the theory of economic decision making according to which economic agents maximize their weighted prospects
RT	"Regret Theory": the theory of economic decision-making according to which economic agents minimize their expected regrets
SH	"Simple Heuristics": the theory of economic decision making according to which economic agents follow a battery of easy-to-apply decision rules

Acknowledgements

Key portions of this book were written while I was a visiting fellow at the Center for the Philosophy of Science at the University of Pittsburgh. The center made for a welcoming and inspiring home away from home, and I am grateful for having had the chance to discuss many aspects of this book with the other members of the center. Without question, this made it a better book. I would especially like to thank Colin Allen, Edouard Machery, and James Justus for their extremely incisive feedback on large parts of this book. Yexin Jessica Li read an early draft of Chapter 6, and provided important criticism that significantly improved that chapter. Audiences at the University of Pittsburgh, the Society for Philosophy and Psychology Annual Conference 2019, the American Philosophical Association Pacific Division meeting 2019, and referees for this press also raised a number of valuable points. Very many thanks to all of these people!

Chapter 2 is a substantially revised and expanded version of my "Niche Construction, Adaptive Preferences, and the Differences between Fitness and Utility" (*Biology and Philosophy*, 2014, 29: 315–335; https://doi.org/10.1007/s10539-014-9439-x). Chapter 3 is a substantially revised and expanded version of my "Firms, Agency, and Evolution" (*Journal of Economic Methodology*, 2016, 23: 57–76; https://www.tandfonline.com). I would also like to acknowledge financial support from the University of Kansas in the form of a General Research Fund grant, which allowed me to spend a summer free from distractions to work on the first chapter of this book. Relatedly, I am very thankful to my department at KU, which made the stay at the Center in Pittsburgh possible.

Finally, I am deeply grateful for all the inspiration, advice, and support that Kelly provides to me on a daily basis. Everything I do is better because of her. James and Elizabeth are anchors that prevent me from floating aimlessly through life, and strong breezes that carry me to distant shores (and sometimes both simultaneously, somehow). I am perpetually thankful for all they do.

Introduction

"Why is economics not an evolutionary science?" asked Thorstein Veblen in 1898 (Veblen, 1898).[1] The answer then, probably, lay at least partly in the fact that both evolutionary biology and economics were quite young subjects, and that it was not clear what insights of relevance could be derived from the former for the latter. In many ways, the situation is now quite different. Researchers increasingly appeal to considerations from evolutionary biology—broadly understood to include all aspects of the study of biological and gene-cultural evolution—to make progress in economics—broadly understood to include all kinds of individual and collective decisions and related phenomena.[2] So, an evolutionary approach has been used to explain the role of emotions in economic decision-making (see e.g. Frank, 1988) and the conflicts between individual and collective interests (see e.g. Frank, 2012; Wilson, Ostrom, et al., 2013). Evolutionary biological considerations are being appealed to in attempts to make sense of the cognitive and neural mechanisms underlying economic decision-making (see e.g. Glimcher et al., 2005; Santos & Rosati, 2015; Smith, 2007), and in the explanation of diversity in human economic decision-making (see e.g. Henrich et al., 2005; Kenrick et al., 2009; Ashraf & Galor, 2013). The understanding of market competition is said to be improved by seeing the latter as an evolutionary process of sorts (see e.g. Schumpeter, 1942; Alchian, 1950; Friedman, 1953; Nelson et al., 2018), and the same goes for economic decision-making itself (see e.g. Okasha, 2011; Orr, 2007; Gintis, 2009; Robson, 2001a, 2001b, 2002; Robson & Samuelson, 2008). As Nelson et al. (2018, pp. 8–9) put it:

> There is good reason to believe that a significant number of empirically oriented economists . . . in fact harbor an implicit evolutionary perspective regarding much of what is going on in the economic world.

In short: implicitly, if not explicitly, economics increasingly *is* an evolutionary science (a remark further supported by the explicitly evolutionary biological perspective of Wilson, Gowdy, et al., 2013).

However, despite this, the methodological foundations of an evolutionary biologically grounded economics still deserve to be further analyzed. In particular, it has not yet been made clear exactly how insights from evolutionary biology can be brought to bear on debates in economics. After all, these two subjects seem to be concerned with quite different sorts of questions. The former is focused on addressing questions such as: Which traits evolve in which environments? Why do they evolve? Which organisms are related to which other organisms to which degree? By contrast, the latter is focused on addressing questions such as: How do consumers and firms decide which goods and services to buy and sell? What determines economic growth? How can we regulate economic activity to the maximum benefit of its participants? On the face of it, it is not exactly clear how these very different sets of questions can be linked to each other.

Relatedly, it is not yet clear what kinds of epistemic progress the application of evolutionary biological considerations in economics makes possible. What *sort* of economic insights can we glean from the appeal to evolutionary biology? When does it allow for major upheavals of debates in economics, when does it merely add minor twists to these debates, and when does it allow for something in between? Answering these methodological and epistemic questions is the aim of this book.

It is clear that doing so is important from the point of view of economics: getting clearer about the nature, promises, and challenges of applying evolutionary biological thinking in economics can help determine where economists can look in order to expand their social scientific toolkit. Given the myriad of challenges we as a species face, and given the importance of economic insights in solving these challenges, an investigation of possible ways of improving or expanding the methodological toolkit of economists is something of much inherent value.

However, the elucidation of the methodological foundations of an evolutionary biologically grounded economics is also interesting from the point of view of the philosophy of science more generally. After all, it is an instance of more a general inquiry into the nature of interdisciplinary research. How can different fields—with different aims, concepts, and methods—be linked to each other? Answering these questions can thus hold lessons for other interdisciplinary fields—including, but not limited to, evolutionary psychology, evolutionary anthropology, and cognitive (neuro-)science. (Indeed, a number of the examples discussed in this book connect straightforwardly to ongoing discussions in these other fields: see especially Chapters 4 and 5.)

To bring out the methodological foundations of an evolutionary biologically grounded economics most clearly, the book begins by showing that there are three different—though related—ways of applying evolutionary biological considerations in economics, each of which has slightly different aims and methods. These forms are the *structural* one, where the relevant processes in economics are shown to be type-identical to some important evolutionary processes; the *evidential* one, where the relevant economic phenomena are shown to be products of some important evolutionary processes; and the *heuristic* one, where evolutionary biological considerations are used as generators for novel and prima facie plausible economic hypotheses. Given this, the book provides an assessment of (i) how these forms function, (ii) what the epistemic benefits they bring, (iii) what challenges they face, and (iv) what specific conclusions they yield.

It is important to stress from the get-go that the book does not aim to provide a comprehensive literature review of all the work that has attempted to bridge evolutionary biology and economics (which would be difficult anyway, given the size of this body of work). Rather, the book starts with making and defending the methodological distinction of structural, evidential, and heuristic connections between evolutionary biology and economics, and then illustrates these connections with a detailed analysis of a small number of specific cases. Of course, a close look at the relevant parts of the literature is still needed here, too. The point is just that the book puts the methodological distinction among the three kinds of evolutionary biologically grounded economics front and center, and does not seek to derive it from a (near) complete review of the field. This methodological distinction structures and drives the discussion of the book as a whole—it is its framework, not its implication. This is also important because there is no existing treatment of the differences and communalities of the three major ways of connecting evolutionary biology and economics in the literature.[3] This is a lacuna that this book tries to fill: for theoretical and pedagogical reasons that will be made clearer in Chapter 1, a thorough understanding of an evolutionary biologically grounded economics greatly profits from keeping this distinction firmly in mind. For this reason, this distinction is best laid out *prior to* engaging with the details of this research program.

Relatedly, it is important to note that the approach the book takes towards the elucidation of the methodological foundations of an evolutionary biologically grounded economics is a heavily practice-based one: it seeks to derive lessons about the way evolutionary biology and economics can be linked by *actually making* some of these linkages.[4] For this reason, the overarching goals of the book are inherently two-sided: they are both to obtain both a clearer view of the merits and challenges of bringing evolutionary biological considerations to bear on economic debates, and a better, evolutionary biologically derived understanding of a number of *specific economic problems*.

With this in mind, the rest of the book is structured as follows. Chapter 1 clarifies and justifies the claim that there are three major ways of connecting evolutionary biology and economics. It lays out the nature of each of these ways and illuminates their interrelations. The remaining chapters are divided into three groups: each of the three forms of an evolutionary biologically grounded economics is being analyzed in one or two chapters, each of which, in turn, focuses on a different application of the relevant form. At the end of each chapter is a list of suggested further readings, so that readers can see for themselves how the issues discussed in the chapter play out in other parts of the literature.

In this way, the book tries to balance breadth and depth. It tries to achieve the necessary breadth by discussing all three of the major ways in which evolutionary biology and economics can be linked. To avoid sacrificing depth in doing this, the book engages in considerable detail with specific questions falling under each of the three ways of connecting evolutionary biology and economics. This kind of structure also ensures that the book retains a strong overall coherence. While each of the six chapters making up its core can, with some minor adjustments, be read as self-standing papers, each of these chapters gains an extra dimension of depth if it is seen as part of the book as a whole. In particular, the contrast in the uses and consequences of the different appeals to evolutionary biology in economics comes out especially clearly if the different chapters are seen as part of "one long argument" (to echo Darwin, 1859).

There are two final points about the book that need to be emphasized from the beginning. First, the fact that the book is focused as it is on five specific cases should not to be taken to imply that there are not many other examples that are not also worth discussing. Rather, this focus is just the result of the fact that it is impossible to discuss everything. Given this, the book concentrates on cases that (a) are particularly conducive to bringing out the methodological lessons to be derived; (b) exhibit the breadth of economic issues that can be analyzed from an evolutionary point of view; and (c) have not been assessed in detail to date.

With regards to point (a), the book thus focuses on cases that help to make particularly clear:

- what needs to happen for there to be structural parallelisms between evolutionary biological and economic processes, and what the epistemic value is of the appeal to such parallelisms;
- what kind of evidential relations between evolutionary biological processes and economic phenomena there are, and what epistemic import they have;
- what heuristic value the appeal to evolutionary biology in economics might have, and how it can function.

With regards to point (b), the cases to be analyzed have been chosen so as to consider many different aspects of economics—from methodological questions about best modeling practices to substantive questions about the nature of economic decision making and collective agency. Indeed, as noted earlier, these cases frequently extend into neighboring fields, like psychology and anthropology, thus making it clear that the issues under discussion go beyond economics per se, and affect the social and cognitive sciences more generally.

Finally, with regards to point (c), the cases have been chosen with a view to ones that have not yet received quite the attention they deserve. The hope is that they will thus be brought to the attention of a wider readership. (Though it is also noteworthy that these cases draw on the work of a number of high-profile scholars in a number of related fields, including Robert Boyd, Sarah Brosnan, Herbert Gintis, Joseph Henrich, Douglas Kenrick, Richard Nelson, Samir Okasha, Peter Richerson, Laurie Santos, and Sidney Winter).

The second point to be emphasized from the beginning is that this is not a work of advocacy. The goal is not, in the first instance, to *show off* what an appeal to evolutionary biology in economics can do.[5] Rather, it is a *critical assessment* of such an appeal: it tries to bring out the promises and challenges of linking the two subjects. Now, since the book tries to achieve this by engaging with the practice of these kinds of linkages, there is a sense in which the book *does* show off what can be done by appealing to evolutionary biological considerations in economics. It is important to keep in mind, though, that this is different from mere advertisement. The aim in actually trying to put evolutionary biological considerations to work in economics is a *balanced assessment* of what can be achieved with such an appeal; importantly, this includes bringing out what *cannot* be achieved in this way.

Indeed, and as also noted earlier, at the most general level, I hope that this book can bring out the value of a particular way of doing philosophy of science. Philosophers of science can be the connectors that help bring different scientific fields together: they are specialist generalists who can help synthesize insights from different sciences. Engaging in this kind of scientific connection-making is an intellectually exciting and valuable venture—and I hope to make a little clearer why that is so.

Notes

1. Marshall (arguably) argued for something similar at that time: see e.g. Stanley (2007).
2. I return to these characterizations of evolutionary biology and economics in the next chapter.
3. As will also be made clearer in Chapter 1, this is not to say that nothing about the way of bringing evolutionary biological considerations to bear in economic debates has been published to date (see e.g. Hodgson & Knudsen, 2010; Hodgson, 2007b, 2011, 2019; Herrmann-Pillath, 2013; see also Campbell, 1965, for a classic take on related issues). The point is just that the specific three-part methodological framework laid out here is novel.
4. The book thus follows the advice of Wimsatt (2007, p. 26): "Don't be afraid to do science, if that seems to be called for to get the necessary perspective!"
5. For books in this vein, see e.g. Nelson et al. (2018), and Hodgson and Knudsen (2010).

1 Three Forms of Evolutionary Economics

I. Introduction

Evolutionary economics, as it is understood here, is a field of research that combines insights from evolutionary biology and economics in a synergistic whole.[1] This very abstract, general understanding of evolutionary economics, while seemingly trivial, is in fact substantive and worth emphasizing from the get-go, as it differs from some other understandings of this field.

In particular, the historically oldest, "classic" understanding of evolutionary economics—going back at least to Veblen (1898), but also having seen famous expositions in Schumpeter (1942), Hayek (1967), Friedman (1953), and Alchian (1950) (among others)—is typically characterized as a form of "heterodox economics" (Vromen, 2001; Hodgson, 1999; but see also Witt, 2008).[2] Historically, evolutionary economists have argued for the rejection and replacement of a number of important, mainstream economic doctrines. Featured prominently among these doctrines are the following (Vromen, 2001; R. Nelson et al., 2018, chap. 1):

(1) Economic agents are only boundedly rational: economic decision-making cannot be analyzed well with models that presume agents are optimizing, computationally unbounded, risk-neutral reasoners.
(2) Competitive markets need to be seen as more than highly efficient transactional spaces, but as engines of innovation for a set of constantly changing firms.
(3) Economic growth is determined not just by capital accumulation or vaguely specified technological advances, but also and largely by the interplay of a complexly intertwined array of entities (such as firms, industries, sectors, countries, and institutions) at many different levels.
(4) Economies are dynamic, nonequilibrium systems and need to be analyzed as such—i.e. they should be studied with the tools of dynamical systems theory (or the like) and not with the kinds of equilibrium analyses that are commonly employed in mainstream economics.

However, while undoubtedly historically important, this relatively narrow, "heterodox" understanding of evolutionary economics will not be central in what follows. There are three reasons for this.

First, as also noted in the introduction, economics in general has become quite pluralistic in its methodology and doctrines, so that it is not so clear that theses (1)–(4) still mark out a unified field separate from mainstream economics (R. Nelson et al., 2018, pp. 8–12). Mainstream economic journals such as the *Journal of Economics*, the *American Economic Review*, or *Journal of Economic Literature* now publish work employing many different theories of choice, many different understandings of the nature of markets and firms, many different approaches towards economic growth, and many different modeling frameworks (see e.g. Bleichrodt & Wakker, 2015; Romer, 1990; Acemoglu et al., 2001; O. J. Blanchard, 1981). Hence, restricting evolutionary economics to theses (1)–(4) makes for a somewhat arbitrary division at best.

Second, the narrower, heterodox understanding of evolutionary economics can anyway be seen as nested in a broader one that just sees the field as combining evolutionary biological and economic considerations with unspecified implications. In particular, while it may well be the case that theses (1)–(4) are true, this need not be seen as a *precommitment* of research in evolutionary economics; it can just be seen as an *implication* of it. For this reason, focusing on a more general form of evolutionary economics can achieve all that the focus on the narrower form can achieve, and (at least potentially) more besides. Note that this is not to deny that theses (1)–(4) are important and deserve to be studied in more detail. Indeed, the specific examples of work in evolutionary economics that make up the bulk of this book all pay special attention to one or another aspect of these core theses traditionally of interest to evolutionary economists. The point is just that there is no reason to see their truth as either to be taken for granted or as required for evolutionary economics to get off the ground in the first place.

Third and finally, a broader understanding of evolutionary economics that sees it as just based on linking—in one way or another—evolutionary biology with economics and which does not make special precommitments to the acceptance or denial of certain economic doctrines has become increasingly popular (Gintis, 2009; Glimcher et al., 2005; Henrich et al., 2005; Ross, 2005; Santos & Rosati, 2015; Kalenscher & van Wingerden, 2011; Y. J. Li et al., in press; Y. J. Li et al., 2012; Chao et al., 2013; D. S. Wilson, Ostrom, et al., 2013; Heller & Mohlin, 2019; L. Samuelson, 2001; A. J. Robson, 2001b). For this reason, there is a lot of current interest in a closer investigation of this broader form of evolutionary economics. This is especially so as this new-found prominence of the broader form of evolutionary economics is not without its detractors either. A number of scholars have claimed that evolutionary economics, ultimately, has little to offer either to economics or to evolutionary biology (see e.g.

Godfrey-Smith, 2009; Sober, 1992; see also Downes, 2018). These critics allege that the contribution of evolutionary biology to economics is either not needed or so weak as to be negligible. Given the fact that many appeals to evolutionary biology in economics have received considerable attention, this controversy thus deserves a closer look.

However, it is also important to note that the present understanding of evolutionary economics, while broader than the traditional heterodox one, is still narrower than some other understandings in the literature.[3] For example, R. Nelson et al. (2018, p. 222) suggest that evolutionary economics should be understood so broadly as to also include—or at least be closely connected to—many other subfields of economics, such as institutional economics or behavioral economics. However, this is not the perspective taken here. While it may well be true that evolutionary economics, institutional economics, and behavioral economics (for example) consider some related phenomena, this does not make them part of the same project. Indeed, what makes evolutionary economics unique—in the present understanding—is precisely the fact that it aims at *combining evolutionary biological and economic considerations*: the outcome of this synthesis, or the set of phenomena considered in it, is not as important as the fact *that* this synthesis is attempted.[4]

With this in the background, the aim of this book can now be stated clearly: bringing out the promises and challenges of combining insights from evolutionary biology and economics in a synergistic whole. What, if anything, can an appeal to considerations from evolutionary biology do for the resolution of open debates in economics—including, but not restricted to, debates surrounding the nature of economic decision-making, market competition, economic institutions, innovation, and non-equilibrium economic systems? What challenges does this appeal face? Can these challenges be overcome? If so, how? Answering these questions is what this book is about.

In order to make progress answering these questions, the first thing that needs to be noted is that, even if it is granted that evolutionary economics seeks to use insights from evolutionary biology to make progress in resolving various open questions in economics, this does not say anything about *how* this is meant to be done. How and why can insights from evolutionary biology be brought to bear on economic questions so as to make progress in resolving these? Answering these methodological questions matters, as it is not possible to assess whether evolutionary economics is successful in linking evolutionary biology and economics until it is made clearer what kinds of linkages are targeted to begin with. (Incidentally, the fact that it is not entirely clear exactly how evolutionary economists try to link the two subjects may also go some way towards explaining why the field continues to be controversial.)

For this reason, this chapter is dedicated to bringing out the nature of the linkages between evolutionary biology and economics. These linkages will then be further explored in detail in the remaining chapters of this book. More concretely, I argue here that evolutionary economics (in the broad version at stake here) comes in three quite different—albeit importantly related—forms: (i) a form that posits and tries to exploit structural parallels between evolutionary biological and economic processes, (ii) a form that tries to exploit evidential links between evolutionary biological phenomena and economic phenomena, and (iii) a form that tries to use heuristic analogies in some of the phenomena investigated in evolutionary biology and those investigated in economics to suggest novel and at least prima facie plausible economic hypotheses to investigate further. While there is much more that needs to be said about the notions of structural parallels, evidential connections, and heuristic analogies, three important points concerning this three-part division in the methodology of evolutionary economics should be noted right away.

First, while this division into three forms of evolutionary economics is novel, it is based on hints and implicit presumptions in the existing literature. Making these hints and implicit presumptions clearer and more precise is one of the goals of the rest of this chapter (and indeed the book as a whole).

Second, while it is true that the three forms of evolutionary economics should be seen as importantly different from each other in their aims and methods, this does not mean that they should be seen as three completely unrelated projects. In fact, there are important relationships among the aims and methods of these three forms (which, of course, is consistent with them being different). Making this clearer is another one of the goals of this chapter (and, again, the book as a whole). For now, though, it is just important to note that the three-part division of evolutionary economics laid out here should not be thought to imply that there are no connections between these three forms whatsoever—i.e. that these are three completely different kinds of projects.

Third, on no form of evolutionary economics is it presumed that evolutionary biological considerations are helpful for *all* economic debates. Rather, the goal is always to just use evolutionary biological considerations to advance *some* debates in economics. This matters, as it implies that, in what follows, evolutionary economics will not be called to task for the fact that there are parts of economics for which an evolutionary biological approach is not useful (see also R. Nelson et al., 2018, pp. 208–210). Instead, the questions to be answered are: are there parts for which such an approach *is* useful? Which parts are those? Why is the evolutionary biological perspective useful in these contexts? How does it achieve its usefulness—i.e. how does the appeal to evolutionary biology work? What are the limitations of the usefulness of the approach? The understanding of evolutionary economics at stake in this book is thus not one that says an evolutionary biological approach will revolutionize all of economics (as has sometimes been claimed for the relationship between evolutionary biology and research in psychology: see e.g. Tooby & Cosmides, 1992), but only that it will *sometimes* be important. However, as I hope to make clearer in the rest of the book, this still makes for an interesting approach towards economics that deserves to be further studied.[5]

The structure of the rest of this chapter is as follows: in sections II–IV, the three forms of evolutionary economics—the structural one, the evidential one, and the heuristic one—are clarified in turn. Section V brings out some of the major relationships between these three different forms, and shows why understanding the communalities and differences of these three forms is important for a thorough understanding of evolutionary economics. Section VI summarizes the discussion.

II. The Structural Form

The key idea behind the structural form of evolutionary economics is that there is a set of processes that are (at least sometimes) operating in *both* the evolutionary biological and the economic realm: forms of "descent with modification." R. Nelson et al. (2018, p. 6) express this as follows:

> [I]n any field of economic activity where innovation is under way, and we argued earlier that in modern economies no field is completely static, there is bound to be a variety of different ways of doing things employed by different actors. At the same time some of these practices, generally but not always ones that are relatively superior in some sense, are expanding in their relative importance, and others, generally relatively ineffective ones, are declining. And as this goes on new modes of operation may enter the picture. This is very much the way traits evolve in biology.

Put differently: evolutionary biology and economics, at least in part, should not be seen as two different subjects, but as two sides of the same subject, namely the study of evolutionary processes—especially selective ones—across generations of entities. To be sure, the entities involved in the two cases are generally different (for example, organisms and genes as opposed to decision options and firms), but this is just a difference in what the processes under study apply to, and not a difference in the nature of these processes themselves. For this reason, some authors have coined the term "generalized Darwinism" as a label for this kind of project: the two disciplines study what are, in effect, shared general (meta-)principles and processes (Hodgson & Knudsen, 2010; Aldrich et al., 2008; Breslin, 2011; Stoelhorst, 2008; see also Beinhocker, 2011). In a bit more detail, this thought can be spelled out as follows.[6]

Consider a population of entities that can, in some way, make copies of themselves or reproduce—i.e. generate new entities to which they bear an ancestral relationship.[7] Assume further that these entities differ from each other in a set of features F. Some of the entities in the population have (some of) the features in F and some do not. Finally, assume that these features are—more or less faithfully—transmitted across the generations of these entities. In that case, we can track the distribution and number of these features among successive time slices of the relevant population. In what follows, I shall call a situation of this kind one involving an *evolutionary process* (Godfrey-Smith, 2009; Sober, 1984; Brandon, 1990). If it is then further the case that the likelihood or ease with which the relevant entities make copies of themselves or reproduce differs depending on whether they have some of the features in F, then this process can be called a *selective evolutionary process* (Godfrey-Smith, 2009; Sober, 1984; Lewontin, 1983; Brandon, 1990).

For present purposes, the key point to note about evolutionary processes is that, as far as their basic nature is concerned, they are neutral as to what kind of entity they apply to (Lewontin, 1970; Godfrey-Smith, 2009; Sober & Wilson, 1998; Okasha, 2006). Put differently, there is nothing inherently "biological" about these kinds of processes; they are not restricted to the entities that feature in contemporary (evolutionary) biology.[8] Rather, at bottom, they are general, quasi-logical/mathematical processes concerning (differential) replication/reproduction (Sober, 1984; Godfrey-Smith, 2009; Dawkins, 1982). To be sure, it could be true that, as a matter of empirical fact, the preconditions for the processes to take place—the existence of entities that can replicate or reproduce, and which differ in some features (which may or may not be relevant to their replicative/reproductive success) that are inherited across generations— only hold in the biological realm. However, this would then be a purely contingent fact: there is no principled, structural reason to see evolutionary processes in general, and selective processes in particular, as restricted to the biological realm. This means that it is at least conceivable that they govern economic entities like decision options or firms just as much as genes or organisms. Nothing rules out the possibility that, apart from genes, organisms, or even groups of organisms, behavioral routines, outlets, or firms can evolve.

However, at this point, a complication needs to be mentioned. This complication concerns the fact that two different (token) phenomena— say, one in biology and one in economics—will differ in many particulars. For example, there might be differences in the mechanisms of inheritance involved in the two phenomena (these might be based on DNA molecules in one case, and pieces of paper in the other), or there might be differences in the speed with which the relevant entities replicate. Given this, when is it the case that two different token phenomena can be seen to instantiate the same *type* of (evolutionary) process?

In response, I suggest that, in order to answer questions like this the individuation schemata of the relevant science are to be used. In this case, this science is evolutionary biology. That is, whether two token processes are to be considered sufficiently similar to be seen as two different instances of an *evolutionary process* is a matter of how the latter kind of process is understood in evolutionary biology. This is not an arbitrary matter of taste, but a question of the scientifically most compelling way of making sense of the relevant set of phenomena. Put differently, the key point to emphasize here is that whether a process in economics is an evolutionary process depends on whether that process has the individuating features of evolutionary processes—as these are understood in contemporary evolutionary biology. (I return to issues related to this point in section V that appears later in this chapter.)[9]

This does not mean that it always needs to be straightforward to determine how a given phenomenon is to be typed. If a process involves Lamarckian inheritance—where acquired features can be inherited and there is a less defined phenotype/genotype distinction—is it still an evolutionary process (see e.g. Boyd & Richerson, 2005; Godfrey-Smith, 2009; Sober, 1992; Jablonka & Lamb, 2005, for more on this)? What if the process involves directed mutations (Sober, 2014)? What if there is no replication, and merely some resemblance between entities in successive generations (Dawkins, 1989; Godfrey-Smith, 2009; Brandon, 1990; Sober, 1984)? Should evolutionary and developmental processes to be distinguished from each—and if so, how strongly (Oyama, 2000; Griffiths & Gray, 1994; Griffiths & Stotz, 2018)? Two points can be noted in response to these questions.

On the one hand, the existence of this kind of controversy certainly needs to be acknowledged (see e.g. Hodgson & Knudsen, 2010). However, this just means that science is subject to change and controversy. While it would be nice to be able to know how, in the ideal limit of inquiry, these questions will be answered, the best we can *actually* do is look towards the best science *currently* and see what it says about how to characterize evolutionary processes. While this may not always lead to clear answers (or any answers at all), it is still the best tool we have.

On the other hand, there is often widespread agreement anyway on how to answer questions like the ones just stated. In particular, it is widely accepted that:[10]

(i) Evolutionary processes do need to involve *reproduction*, rather than just differential survival.

(Brandon, 1990; but see also Godfrey-Smith, 2009)[11]

(ii) Evolutionary processes do not need to involve *replication*, but merely reproduction with some resemblance.

(Godfrey-Smith, 2009)[12]

(iii) Evolutionary processes can feature a variety of inheritance or mutational processes.

(Boyd & Richerson, 2005; Godfrey-Smith, 2009; Sober, 2014)

While these sorts of widely accepted claims need not resolve all questions about how to type processes in evolutionary biology and economics, they are sufficient for present purposes. In particular, the rest of this book can just rely on features (i)–(iii) to be the key elements of the standard characterizations of evolutionary and selective processes; nothing of what follows depends on more controversial aspects of the characterization of these processes.

With all of this in mind, the epistemology of the structural form of evolutionary economics can be made clearer. If the same sort of process governs evolutionary biological and economic entities, then all that we have learned about the way evolutionary and selective processes work in evolutionary biology can be straightforwardly transferred to at least some economic phenomena.[13] Since the reason why we can link evolutionary biology and economics, in the structural form, is that these are at bottom the same subject (at least in part), this means that insights concerning the key features of evolutionary processes derived in one context also have to apply in the other. The best way to make this clearer is by briefly considering two examples of this form of evolutionary economics. These examples will be discussed in more detail in Chapters 2 and 3; in the present context, they are just to be seen as illustrations of this form of evolutionary economics.

First, note that one of the core processes of economics—individual economic decision-making—can look a lot like a selective evolutionary process. The agent begins with a set of options (different courses of actions with differently preferred consequences) and ends up choosing the option with the highest preference value. (As will be made clearer in Chapters 2, 4, and 5, things are a lot more complex than this when analyzed in more detail; but for now, this gloss is sufficient for illustrative purposes.) This seems to be parallel to evolution by natural selection, where a population in an environment begins with a set of options (different traits with differing fitness consequences) and ends up coalescing on the option with the highest fitness. (As will also be made clearer in Chapter 2, things are more complex here too when analyzed in more detail; but again, this gloss is sufficient for illustrative purposes.) Moreover, it may even appear that fitness and preference values are at least quite often correlated: what economic agents prefer often tends to be what is biologically good for them. (Once more, Chapter 2 will bring out the complexities of this assumption in more detail.) For these reasons, one might propose that economic decision-making is structurally parallel to evolution by natural selection: at bottom, these are different instances of the same constrained optimization process (for versions of a view like this, see e.g. Grafen, 1999, 2006; Okasha, 2011; Okasha, 2007; Sterelny, 2012b; see also Gintis, 2009; A. J. Robson, 1996; Orr, 2007).[14]

If this is the case, it will be important, as it will allow the transfer of key theoretical insights from evolutionary biology to economics. For example, evolutionary biologists have long noted that low-risk strategies can be fitness maximizing relative to high-risk ones (Gillespie, 1977; see also Okasha, 2007; Schulz, 2008). If this insight can be transferred into economics, this would be useful, as it would show that risk-aversion should be seen as a core feature of economic decision-making. In turn, this would help clarify important debates in economics, where the relationship between attitudes towards risk and economic decision-making continues to be a point of contention (Okasha, 2007; Schulz, 2008; Cooper, 1987, 1989; Orr, 2007; Winterhaler, 2007; A. J. Robson, 1996).

However, individual economic decision-making is not the only economic process that the biological process of natural selection may be thought to be parallel; another major such process is market competition. Specifically, several researchers have suggested that competitive markets can function as selective environments in the same way that biological environments can function as selective environments (see e.g. Friedman, 1953; Alchian, 1950; R. Nelson & Winter, 1982; R. Nelson et al., 2018). So, it appears that firms can (a) reproduce (i.e. create offspring firms), (b) differ in a number of features (such as the extent to which they have a cooperative firm culture) which (c) are transmitted across generations of firms, and that (perhaps) (d) affect the likelihood and manner in which these firms can reproduce themselves (e.g. through their effect on firm profitability). This is important, as (a)–(d), as just noted, are also exactly the conditions that underlie evolution by natural selection (Godfrey-Smith, 2009; Sober, 1984; Brandon, 1990).

If so, then this would again allow us to transfer insights from one of the sciences to the other. One of these insights concerns the fact that there are situations where "collectives"—beehives, multicellular organisms, or mutualistic arrangements between plants and fungi (mycorrhiza)—are selected for and thus evolve *as a unit* (Maynard Smith & Szethmary, 1995; Godfrey-Smith, 2009; Okasha, 2006; Clarke, 2016). If the existence of a natural selection/firm competition parallelism can be underwritten with compelling arguments, therefore, this would allow an inference to the conditions under which firms composed of many different employees would also be selected (by a competitive market) *as a collective unit*. In turn, this would be important, as it can help clarify what firms *are*: are they economic agents of their own or mere transactional spaces for other economic entities to interact in? Since the latter is still a hotly debated issue in economics (Hodgson, 1999, pp. 247–249; Hodgson & Knudsen, 2010; R. Nelson et al., 2018; List & Pettit, 2011), evolutionary biological support in resolving it can be highly useful.

The details of both of these examples is the topic of Chapters 2 and 3 that follow. What matters for now is just that they illustrate the nature of the structural form of evolutionary economics: this type of evolutionary economics is based on the idea that there are structural similarities between evolutionary biology and different parts of economics—from consumer theory to producer theory—and allow for the transfer of insights from one subject to the other.[15]

III. The Evidential Form

The second form of evolutionary economics is evidential in nature. The key idea behind this form is that key parts of economics concern the study of decision-making (especially when it comes to decisions among scarce resources or within informational constraints), and that evolutionary biological considerations can be helpful in advancing this study. Santos and Rosati (2015, p. 323) express this point as follows:

> [T]he decisions that humans face in economic contexts often have clear analogues with the problems that animals face when foraging for food or seeking mates. Consequently, the types of choices that psychologists and behavioral economists focus on in humans—such as intertemporal preferences and risk preferences—are also ubiquitous in biology and behavioral ecology. . . . [W]e argue that comparative research on decision-making biases in primates is critical for understanding the decision-making biases observed in humans.

Similarly, Kenrick et al. (2009, p. 764) say:

> Drawing on an evolutionary perspective, we propose that people make decisions according to a set of principles that may not appear to make sense at the superficial level, but that demonstrate rationality at a deeper evolutionary level. By this, we mean that people use adaptive domain-specific decision-rules that, on average, would have resulted in fitness benefits. Using this framework, we re-examine several economic principles. We suggest that traditional psychological functions governing risk aversion, discounting of future benefits, and budget allocations to multiple goods, for example, vary in predictable ways as a function of the underlying motive of the decision-maker and individual differences linked to evolved life-history strategies.

Before considering this form of evolutionary economics in more detail, it is important to note two points about the understanding of economics that underlies it.

First, it may be thought that the study of individual (economic) decision-making is not, in fact, a form of economics at all—at least if this study is understood in a descriptive, mentalistic manner. In particular, it may be thought that economics studies either just choice *behavior*—i.e. the *results* of a mental process of (economic) decision-making—or the *rationality* of different forms of economic decision-making—i.e. the kinds of decisions people *ought* to make (for the former, see e.g. Gul & Pesendorfer, 2008;

Friedman, 1953; for the latter, see e.g. Joyce, 1999; Bradley, 2017). The study of the mental mechanisms underlying (economic) decision-making, by contrast, should just be thought to be a form of psychology or neuroscience (or a related subject).[16] Hence, it may be thought that this form of evolutionary economics is not actually a form of evolutionary economics at all—it is really a form of evolutionary psychology.

However, as will also be made clearer in Chapter 4, this conclusion is overly strong. In the first place, the academic integrity of economics does not depend on it being cleanly distinguishable from cognitive science, social psychology, or neuroscience (though see also Hausman, 1992). It is now widely recognized that many disciplines differ from each other only in emphasis, not in subject matter. For example, the differences between linguistics and psychology (see e.g. Branigan & Pickering, 2017), psychology and neuroscience (see e.g. Gazzaniga et al., 2009), and biology and chemistry (see e.g. Tan, 2005), are now widely recognized not to be hard and fast. The same is true for economics, psychology, and neuroscience: many scholars now see these as investigating a common set of topics (see e.g. Thaler, 2000; Glimcher et al., 2005; Camerer, 2007).

Furthermore, even if the prime goal of economics is seen as the prediction or normative evaluation of choice behavior, the psychological underpinnings of the ways in which people make economic decisions would still be useful as an input into both of these projects. After all, in order to know how people will choose, it is useful to know the processes leading up to these choices (Hausman, 2008), and in order to provide a plausible account of the decisions they ought to make, it is useful to know something about the kinds of decisions they actually do make (Lyons et al., 1992). For these reasons, the study of economic decision-making should indeed be seen to be a key part of economics (perhaps *alongside* the prediction of choice behavior and the normative evaluations of people's choices).

Second (and relatedly), it also needs to be acknowledged that there is more to economics than the study of individual decision-making. For example, the theory of the firm, welfare economics, international economics, and general equilibrium analysis are all also crucial parts of economics, but they do not (directly) concern individual decision-making.[17] Given this, it may appear that the focus on individual decision-making that is characterizing the evidential form of evolutionary economics is overly restrictive.

However, this focus on individual decision-making does not in fact pose a problem for the evidential form of evolutionary economics. This is due to the moderate ambitions of evolutionary economics in general (at least how this field is understood here). As noted earlier, I do not presume that evolutionary economics needs to be relevant to every part of economics—this would be overly strong and unconvincing. As long as evolutionary economics can make a contribution to *some* parts of economics, that is all that is necessary. If it is therefore granted—which, as just noted, is reasonable—that the study of individual economic decision-making is *a* (key) part of economics, then that is all that is needed in order to get the evidential form of evolutionary economics off the ground.

With all of this in mind, the core idea behind the evidential version of evolutionary economics is that evidence about the ways in which humans make economic decisions can be obtained from the consideration of certain evolutionary biological findings. In the background of this appeal to evolutionary biology in economics are the ideas that (a) human economic agents should be seen as evolved biological entities and (b) understanding them as such is useful for understanding how they make economic decisions (see also Collins et al., 2016; V. Smith, 2007; Hammerstein & Stevens, 2012). More specifically, there are two different ways in which the evolutionary biological evidence to underwrite this form of evolutionary economics can be obtained; these two ways roughly correspond to the two core principles of evolutionary theory—the principle of common ancestry and the principle of evolution by natural selection (Sober, 2011).

First, the evidence from evolutionary biology could be derived from phylogenetic studies. If economic decision-making is seen as a biological trait, then, given the fact that humans are relatively closely related to many other organisms (and other primates in particular), it is plausible that we can learn something about human economic decision-making by studying economic decision-making in other animals. (This is in line with the quote from Santos & Rosati, 2015 in the previous section.). Again, a concrete example will make this clearer; the example (and related other ones) will be analyzed in more detail in Chapter 4.

Several researchers have used single-neuron monitoring to assess the ways in which macaques make decisions in both certain and risk-involving environments, and both in the context of individual and strategic decisions (see e.g. Glimcher et al., 2005; Platt & Glimcher, 1999; Santos & Chen, 2009). What they have found is that neurons in the macaque brain seem to encode the relative desirabilities of the different actions available to the organism, and that the macaque decides what to do by assessing which of these desirabilities is highest (Glimcher et al., 2005). In short, at least on a neural level, macaques seem to make at least some decisions in an optimizing manner, much as it is set out in the classical theories of economic choice. For present purposes, the key point to note about this research is that the conclusions it is trying to reach are not restricted to macaque decision-making (though that is of course interesting, too). Rather, this research also tries to tell us something about *human* economic decision-making (Glimcher et al., 2005; Santos & Chen, 2009). In particular, these findings are meant to make it plausible that humans, too, at least sometimes make decisions (at least on a neural level) by maximizing their expected desirabilities (Glimcher et al., 2005). To make this inference, these researchers rely on the fact that macaques and humans are relatively closely related, and that many parts of the macaque brain have homologues in the human brain (Glimcher et al., 2005; Santos & Rosati, 2015).

Now, as will be made clearer in Chapter 4, this step from non-human to human decision-making needs much closer scrutiny, and there are also many other findings that need to be taken into account here. For present purposes, though, the key point is just that this illustrates quite clearly one of the ways that the evidential form of evolutionary economics operates: namely, by relying on the fact that human and non-human animals are part of the same tree of life.

The other main way in which this form of evolutionary economics operates is by relying on studies about the selective or other evolutionary pressures on human decision-making mechanisms directly. That is, rather than relying on comparative studies (i.e. by considering the decision-making of related organisms), evolutionary economists can also assess the evolutionary history of human economic decision-making in and of itself.[18]

This appeal to the evolutionary history of human economic decision-making can be informative for a number of different reasons. One of the major reasons is that it can yield a better understanding of the nature of the *diversity* that has been in found in the ways that humans make economic decisions. In particular, while it has become increasingly clear that different groups of humans—e.g. different genders or different cultures—make economic decisions in different ways (Byrnes et al., 1999; Henrich et al., 2001; Henrich et al., 2010), it is not yet clear whether these differences are "deep" in structure, or whether they are merely superficial. That is, it is not clear whether (a) at a fundamental level, all humans make economic decisions in the same way, with any differences in the resulting choice behaviors being merely due to the fact that the "inputs" into these same decision-making processes differ across humans; or whether (b) different humans rely on fundamentally different decision-making mechanisms.

Knowing more about the evolutionary history of humans could aid in the investigation of this issue. In particular, the appeal to what is known about human evolutionary history could tell us something about whether there were divergent cultural or genetic evolutionary pressures on human decision-making mechanisms in different groups of humans, or whether there is a set of evolutionary pressures that all humans have been subject to that has led to the mechanisms underlying human economic decision-making becoming human universals. (This is in line with the quote from Kenrick et al., 2009.) Naturally, it is also possible that the former explanation holds for some aspects of the existing diversity in human economic decision-making, and the latter for others.

Of course, whether it is *in fact* the case that consideration of what is known about human evolutionary history could tell us something about the ways in which humans make economic decisions requires much closer scrutiny; Chapter 5 is dedicated to doing this. For now, though, the previous example is sufficient to illustrate the fact that an appeal to findings about the evolutionary history of the human species could provide a deeper understanding of the nature of the diversity in human economic decision-making.

Therefore, overall, at the heart of the evidential form of evolutionary economics is an appeal to the fact that (human) economic agents are biological systems that have been shaped by the same forces that have shaped other biological systems. In turn, this is thought to be able to advance our understanding of the ways in which they make decisions.

It is important to note that the evidential version of evolutionary economics merely seeks to use evolutionary biological considerations as *evidence* for settling disputes in economics (hence the name). That is, the evolutionary biological considerations are merely meant to *underwrite*—provide reasons in favor of—hypotheses in economics (for more on the notion of evidence, see e.g. Sober, 2008; Achinstein, 2013). I return to this point in section V that follows, but for now, it is sufficient to note that this sense of "underwrite" can be formally spelled out in many different ways, for example, in terms of likelihood ratios (Goodman & Royall, 1988; Royall, 1997), severity of tests (Mayo, 1996), decreases in imprecision (Joyce, 2010), or much else as well (Sober, 2008; Achinstein, 2013). For present purposes, though, the key thing to note about this relation is that it is "local": A can be evidence for B, even though B is false (or did not occur, etc.). In fact, A can be evidence for B even though we have very good reason for thinking that B is false (or did not occur, or the like); in particular, we may have much other evidence (C, D, E, . . .) for A being false (or not occurring, etc.) (Sober, 2008). The point is just that A provides *a reason* for thinking that B is true (occurred, etc.).

In the present context, this implies that the evidential form of evolutionary economics should be seen to use considerations from evolutionary biology to make a given economic theory more or less plausible. This does not (or at least need not) mean that these evolutionary biological considerations are used to force us to *accept as true* the relevant economic theories; the claim is just that these considerations can increase or decrease our commitment to the relevant economic theories (see also Schulz, 2018b).[19] In short, the key idea behind the evidential form of evolutionary economics is that evolutionary biological considerations can present a reason for thinking some given economic theory is true (or false)—nothing more, but also nothing less.

IV. The Heuristic Form

The third and final form of evolutionary economics is purely heuristic in nature. The core idea behind this form is that there are some similarities in the topics studied, the approaches used, and challenges faced by evolutionary biologists and economists. While these similarities need not amount to a genuine parallelism—as in the structural form—and while they need not provide genuine reasons for thinking that a particular hypothesis is true—as in the evidential form—they can still be useful for suggesting interesting and novel hypotheses to consider further.[20] Galla and Farmer (2013, p. 1233, emphasis added) express this point as follows:

> An example of work in a similar spirit [to Galla & Farmer's economic model] is the 1972 paper of Robert May . . . which analyzed the generic stability properties of differential equations modeling predator—prey interactions with random coupling coefficients. This work challenged the conventional wisdom that more complex ecosystems are necessarily more stable than simple ones by showing that for a particular ensemble of random equations the opposite is true. The question of whether complex ecosystems are more or less stable remains controversial. In any case, May's paper has played a vital role by *focusing the debate* and *forcing ecologists to think carefully* about the generic properties of ecological interactions. *Our intent is similar.*

In a bit more detail, this idea can be spelled out as follows.

At its most general, a heuristic is something that functions to efficiently accomplish a given task, but which does so in biased manner and with a specific error profile (Wimsatt, 2007, appendix B; Simon, 1957; Gigerenzer et al., 2000). In the present context, though, I understand "heuristic" more narrowly. In particular, I here take a (scientific) heuristic to be something that *suggests novel and at least prima facie plausible hypotheses* (see also Schulz, 2011b, 2015, 2012). That is, in the way I understand them here, (scientific) heuristics are tools with a specific function—namely, suggesting plausible, novel hypotheses. While this is obviously a more narrow understanding of the term than the one employed by some other authors (such as Wimsatt, 2007), it is not without precedent, and is in fact in keeping with a fairly common usage of the term (Machery, forthcoming; Machery & Barrett, 2006; Hesse, 1963; Hausman, 1992, p. 99; Godfrey-Smith, 2009, p. 65). A few further points need to be said about this relatively narrow characterization of heuristics.

First, while narrow, the present characterization leaves open exactly what the *source* is of the suggestion of novel, prima facie plausible hypotheses. This could be the consideration of a given evolutionary biological phenomenon, a given evolutionary biological fact, a given evolutionary biological hypothesis, or much else. The point is just that the consideration of the evolutionary biological item in question suggests economic hypotheses that have not been considered before, and which are at least prima facie plausible—where "economic hypotheses" are simply hypotheses relevant to a particular economic debate or question. As understood here, then, the appeal to considerations from evolutionary biology in economics can be defended from a heuristic point of view to the extent that this appeal suggests novel and prima facie plausible hypotheses about a given economic debate or question.

Second, I do not assume that the appeal to a particular evolutionary biological consideration needs to be able to *continuously* suggest novel and plausible hypotheses. All that is required is that this appeal can do so at least once. The motivation behind this is that there is value in the *single* suggestion of novel and plausible economic hypotheses. Since the latter are always needed and sought for, every little bit helps (see also Machery, forthcoming). At any rate, from the point of view of the study of evolutionary economics more generally, the key point is that the appeal to evolutionary biological considerations in economics can be defended from a heuristic point of view if some appeals to evolutionary biological considerations can suggest novel and plausible economic hypotheses— even if each of these appeals is heuristically useful only once.

The third point about the present characterization of heuristics that needs to be made clearer concerns the need for the suggested economic hypotheses to be novel and plausible. How are these terms to be understood, and why do I restrict the heuristic use of evolutionary biological considerations in economics to the suggestion of economic hypotheses with these features?

Under *novel*, I understand hypotheses (or aspects thereof) that have not been considered up to now, and that would not have been considered but for the consideration of the heuristic at stake. So perhaps the hypotheses in question have never seriously been considered at all, or perhaps they have not been given the attention that they deserve—in which case the novelty of the heuristic lies in the *importance* that needs to be given to them, or perhaps there is an aspect of the hypotheses that has been overlooked thus far—in which case the novelty lies in this overlooked aspect. That is, for something to be a useful heuristic in the sense that is relevant here, it has to actually suggest something *new*. The interest in the present context lies in the possibility to use considerations from evolutionary biology to advance economics by suggesting new ways of thinking about some economic question or phenomenon. In the words of Galla and Farmer (2013) from the previous quote, the goal of interest here is "focusing the [relevant] debate" and "forcing [economists] to think carefully about" the relevant issues. This is not to say that the suggestion of something already known may not be important for other reasons (Worrall, 1989; Hitchcock & Sober, 2004). It is just that *novelty* is a key part of the heuristic form of evolutionary economics *as that is understood here*.

Now, it needs to be acknowledged that it need not be obvious to what extent a given hypothesis is in fact novel. Is it enough that the relevant hypothesis is new to a given researcher? Does it have to be new to the majority of researchers in the field? To all of them? Fortunately, it is not necessary to settle these questions here. The heuristic usefulness of an appeal to evolutionary biological considerations in economics can be seen to be a matter of degree: the less widely known the relevant hypothesis is, the more useful the appeal to the evolutionary biological considerations can be seen to be (also keeping mind that there can be differences in exactly what the novelty consists in to begin with). So, there is no need to specify an exact degree of novelty to spell out the heuristic form of evolutionary economics; it can be acknowledged that the plausibility of an application of this form of evolutionary economics depends on the details of *how* novel the suggested hypothesis is.

Concerning the *plausibility* of a suggested economic hypothesis, the following can be said. Under "plausibility," I understand hypotheses that are not known to be inconsistent with other findings, and that open up new avenues for investigation. The suggestion of hypotheses that are known to be inconsistent with accepted findings (either about the phenomenon at issue or other phenomena), while novel, is not very useful, as investigating these hypotheses further is a known dead-end. Note that the inconsistency in question needs to be *known*: if it is not known, then the suggestion of a given hypothesis that is, in fact, inconsistent with other accepted theories can still be interesting. After all, it can be very fruitful to determine *that* a given hypothesis is inconsistent with known theories. What is to be ruled out is just the suggestion of hypotheses that are known to be non-starters.

It is especially this second feature that needs to be seen as a key aspect of the heuristic form of evolutionary economics. There is little usefulness in a mere *expansion* of the hypothesis set under consideration. The problem in economics (and all other sciences) has never been the fact that we are running out of hypotheses to test. There is no shortage of novel hypotheses we *could* consider. Take any known hypothesis, and add an arbitrary parameter to it ("people make decisions by maximizing the sum of expected utility *and the number of crayons in their possession*"), or take any known hypothesis, and change it in an arbitrary way ("people make decisions by *minimizing* expected utility"). Coming up with *something new* is not the problem. The problem is (and has always been) the fact that we need *plausible* hypotheses to test. If we do not know how to explain a given phenomenon, the issue is coming up with hypotheses that might explain it—not just *any* novel hypothesis. It is easy to come up with novel hypotheses. It is hard to come up with novel hypotheses that *might actually explain* the phenomenon of interest, and that cannot be brushed aside as immediate non-starters.

However, it is also important not to go too far in the opposite direction—the fact that a hypothesis is plausible does not mean it is true. In fact, by itself, the fact that a hypothesis is plausible does not even provide *a reason* for thinking the hypothesis is true. A plausible hypothesis, as that is understood here, is one that is not *ruled out* by what is known about the phenomenon at issue; it is not one for which we have a positive reason for thinking it is true. While, as just noted, this is not trivial, it does fall short of actually providing support to the hypotheses suggested.[21]

The final point that needs to be noted about heuristic devices as they are understood here is that, on one level, there is something inherently "observer-dependent" in the nature of heuristic devices. What makes the consideration of a certain evolutionary biological phenomenon heuristically useful is that, at least for human observers, (certain aspects of) this phenomenon bring(s) to mind novel and plausible hypotheses concerning a given economic phenomenon or question. This does not necessarily generalize to other cognitive systems—e.g. artificial ones—and may depend inherently on the particular cognitive biases of human minds.[22] However, far from being problematic, this kind of observer-dependence is in fact a key feature of heuristic devices, and needs to be embraced as such (Wimsatt, 2007).

That said, it is important to note that the observer-dependence of heuristic devices does not mean that there is no way to further underwrite or explain their use. On the one hand, a plausible heuristic use of an appeal to evolutionary biology in economics requires that it *actually produces* novel and plausible economic hypotheses. Just saying that it *might* do so is not enough—the hypotheses need to actually be produced. The claim that a given evolutionary biological phenomenon *might* be able to produce novel and plausible economic hypotheses is too cheap epistemically. Arguably, this can be said about nearly every evolutionary biological phenomenon. Plausible defenses of the heuristic usefulness of the appeal to an evolutionary biological phenomenon in economics will point to concrete novel and prima facie plausible hypotheses that have *in fact* been suggested by the appeal to the evolutionary biological phenomenon.

On the other hand, the reason why a given evolutionary biological phenomenon can suggest novel and plausible hypotheses about a given economic phenomenon need not in any way be mysterious. In particular, a common way in which the consideration of a phenomenon in evolutionary biology can be heuristically useful in economics is when the two cases share a number of features that, while falling short of making them tokens of the same type of phenomenon, still mark them out as relatively similar. I return to this point in the next section (as well as in Chapter 6). For now, it is just important to stress that, while heuristic usefulness is clearly a mind- and language-dependent phenomena in some respects, this does not mean that there is nothing to say about whether something is a good heuristic device or not. To make all of this clearer, it is again useful to briefly consider an example of this form of evolutionary economics in action (this example will be analyzed in more detail in Chapter 6).

Consider the question of how compelling equilibrium modeling is as an investigative strategy in economics. As noted earlier, a number of researchers (see e.g. Carlaw & Lipsey, 2012; R. Nelson & Winter, 2002; R. Nelson et al., 2018, p. 19; Galla & Farmer, 2013) have come to question economists' heavy reliance on the analysis of equilibrium models—i.e. models that eventually settle into some kind of stable state, and where the nature of this stable state is the focus of the analysis of the relevant system. These critics are concerned that real economic systems are often dynamic in nature and virtually never in any kind of equilibrium, and that the reliance on equilibrium modeling therefore misses many of the key features of many of these systems.

Now, it turns out that, due to the many different issues in play, assessing whether and when this criticism hits the mark is quite complex. Fortunately, it appears that an appeal to considerations from evolutionary biology might help here. In particular, in line with the previous quote from Galla and Farmer (2013), a very similar debate to the one in economics has been going on in evolutionary ecology. There, too, equilibrium modeling is quite common, but critics have come to question the plausibility of this approach (R. May, 1974, 1972). In turn, this suggests that it may be possible to learn something about the former debate by looking at the latter one in more detail. Specifically, there is reason to think that it may be fruitful to consider the idea—floated in the context of evolutionary ecology (R. May, 1974)—that the plausibility of equilibrium modeling depends on the extent to which the systems to be modeled are "sorted." If non-equilibrium systems are more likely to disintegrate, then it is plausible that, among the systems that we are actually interested in modeling—i.e. which have a minimum level of stability—a significant proportion are equilibrium systems. Importantly, though, this suggestion is just that: a suggestion. The fact—if it even is a fact—that ecosystems are often equilibrium systems because they are sorted systems does not provide a structural or evidential reason to think that economic systems are too (or the opposite). However, this does not mean that this suggestion is not interesting; it just means that this suggestion still needs to be further explored.

Therefore, the heuristic form of evolutionary economics is based on using evolutionary biological considerations to suggest novel and plausible economic hypotheses to consider further. While there is no guarantee or reason to think that that these hypotheses will in fact turn out to be true, it is at least the case that their consideration moves economics forward.

With this, the initial statement of the three forms of evolutionary economics is complete. What needs to be done now is to consider how these three forms are related to each other.

V. The Relationships Between the Three Forms of Evolutionary Economics

The fact that evolutionary economics splits into a structural form, an evidential form, and a heuristic form might suggest that it is a very "disunified" field. Indeed, on the face of it, it may seem that this is not *one* project, but *three*, quite different projects. In turn, this might then raise the question why all three forms of evolutionary economics should be discussed in the same book. Is the fact that all three of them try to combine evolutionary biology and economics really sufficient to consider them part of the same overarching project? Would it not be more convincing and less confusing if each of these three forms evolutionary economics are discussed separately from the others (in their own books, perhaps)?

However, as I make clearer in this section, these worries are unfounded. There are in fact a number of important connections among the three forms of evolutionary economics. More importantly, as I further try to show in the remaining pages of this chapter (and indeed the book as a whole) understanding these connections is very useful for a thorough understanding both of the field as a whole and of each of the individual forms of evolutionary economics separately. To see this, begin by considering the major aspects of the relationships between the three different forms of evolutionary economics. After that, consider the reasons why an understanding of these relationships is crucial for a thorough understanding of the nature of a (broad form of) evolutionary economics.

1. Connections Between the Structural and the Evidential Forms

The evidential form of evolutionary economics differs from the structural one in that the latter sees economics as studying *evolving* systems (such as individual economic decisions or competitive markets). By contrast, the evidential form sees economics as studying a particular kind of *evolved* system—namely, human agents. So, on the latter form, it is not assumed that we need to see evolutionary biology and economics as two sides of the same subject (at least in part); rather, here, all that we need to acknowledge is that economics deals with entities that have an evolutionary history that it is useful to take into account when studying their interrelations.

Note that technically it is of course true that the existence of a structural parallelism between an evolutionary biological and economic phenomenon can be seen as a reason—i.e. as evidence—to favor some particular hypothesis concerning that phenomenon. However, the reverse is not true: not all reasons to favor some particular economic phenomenon need to be based on the existence of a structural parallelism. In particular, we can obtain evidence about the nature of one type of phenomenon from the consideration of another type, as long as these two phenomena stand in the right kinds of relations. Exactly which kinds of relations these are is not entirely clear (see e.g. Achinstein, 2013; Sober, 2008), but it is clear that *one such* relation concerns that between processes and their products—or, more generally, that between cause and effect (Sober, 2008; Dretske, 1981). It is this kind of relation that will be at the forefront here.

Given that the core difference between the structural and evidential forms of evolutionary economics thus consists in the fact that the former focuses on cases where the same processes are occurring in evolutionary biology and economics, whereas the latter focuses on cases where economic processes are products of evolutionary biological processes, it becomes clear that the former is (generally) epistemically weaker than the latter. The structural form of evolutionary economics allows the wholesale transfer of insights from evolutionary biology to economics. In the relevant respects, these turn out to be essentially the same subject. By contrast, the evidential form of evolutionary economics just provides *prima facie reasons* for supporting some economic theories—we are trying to infer something about the nature of a phenomena by considering the processes that produced that phenomenon. For this reason, the evolutionary biological considerations do not directly translate into economic insights—i.e. provide the strongest possible reasons for the truth of some economic hypothesis—but only provide *some* support for various economic theories.

That said, though, the provision of reasons—evidence—is still an important part of science (Achinstein, 2013; Sober, 2008; Goodman & Royall, 1988; Schulz, 2018b, chap. 3). While it may not be able to fully underwrite an economic hypothesis, it can still add key pieces of support to this hypothesis. This is especially useful when other pieces of support are lacking, ambiguous, or not very strong. Often, all that we can do when investigating some particular problem is to assemble pieces of evidence. While each of these pieces may be weak and uncertain in their own right, together they can make for a compelling picture of what the best solution to the problem in question is. Importantly also, the value of a piece of evidence is not reduced by the fact that it is not yet clear what the other pieces are like, or how they can be integrated. A genuine reason in favor of a hypothesis remains such a reason—even if it ends up being outweighed by other such reasons.[23] In this way, the evidential form of evolutionary economics can still play an important part in the investigation of economic questions, though it is epistemically weaker than the structural form.

2. Connections Between the Evidential and the Heuristic Forms

In one sense, the heuristic form of evolutionary economics can be seen as a weakened version of the evidential form. In particular, evidence can be a heuristic: if the consideration of one phenomenon P1 provides a reason for thinking that a hypothesis H about a different phenomenon P2 is true, then if H is novel, the consideration of P1 can also (trivially) be seen as heuristically suggesting that H *might* be true about P2. (In fact, this sort of heuristic connection is especially plausible for very weak evidential relations.) In this sense, the heuristic form of evolutionary economics can be seen as a modally weakened version of the evidential form.

However, in another sense, the two are more strongly independent of each other. In particular, the consideration of one phenomenon can be heuristically useful for the investigation of another without being evidence for it. This is due to the fact that the heuristic usefulness of the consideration of phenomenon P1 for the understanding of phenomenon P2 may rest in nothing other than the fact that that P1 and P2 have a number of peripheral features in common which suggest—to (some) human researchers—some novel and plausible hypotheses about P2. These features need not involve cause/effect relations (or other evidence-grounding relations) but might merely be based on some similarities that, to the relevant human observers, are suggestive.

This is illustrated well by the discussion of the usefulness of equilibrium modeling in economics. As was hinted at in the previous section, and as will be made clearer in Chapter 6, the idea that equilibrium modeling is successful in evolutionary ecology due to the fact that (many) of the systems to be modeled are sorted systems suggests that something similar may be true in economics. However, the reasons for which economic systems are sorted systems (if they are sorted systems at all) turn out to be quite different from the reasons for why ecosystems may be sorted systems. Hence, it is just not plausible to see the (alleged) fact that equilibrium modeling in evolutionary ecology is successful because ecosystems tend to be sorted systems as providing *evidence* for the fact that something similar is true in economics. These two cases are just too different for this to be true. The consideration of the situation in evolutionary ecology provides no reason to think that economic systems will be sorted systems that can be modeled well using equilibrium models, though it does *suggest* that *exploring* this further is useful.

It is worthwhile to emphasize that while the heuristic form of evolutionary economics is the epistemically weakest form of the three kinds of evolutionary economics, this does not mean that it is not still epistemically interesting and important. Particularly in contexts in which economists are very unsure about the proper way of proceeding—such as those involving questions that are not easily amenable to direct, empirical analysis, but concern highly theoretical or abstract matters (such as the reasonableness of equilibrium modeling)—the suggestion of novel and plausible hypotheses can be very useful. Hence, it is important to keep in mind that while the heuristic form of evolutionary economics need not *directly* deepen our understanding of some economic phenomenon, it can *indirectly* do so. It can point to ideas worth considering further that in turn might lead to a deepening of our understanding of some economic phenomenon.

3. Connections Between the Structural and Heuristic Forms

The third and last set of connections between the different forms of evolutionary economics worth discussing occurs between the structural and the heuristic forms. Several aspects of this connection were already mentioned when laying out these two forms individually, but it is worthwhile to bring them together here.

In particular, as noted in the discussion of the structural form of evolutionary economics, whether two processes are considered to be fundamentally of the same type depends on how these processes are individuated in the relevant sciences. This matters, as for a structural parallelism—and thus the structural form of evolutionary economics—*some similarity* between two phenomena is not sufficient. The similarity needs to concern the scientifically recognized individuating features of the type of phenomenon in question. By contrast, for a merely heuristic analogy between the two phenomena, merely "peripheral" similarity between the two phenomena is sufficient. In fact, as just noted, this is one key way in which heuristic connections among two different phenomena can be justified: two phenomena can be seen as falling under distinct types, but, on the currently accepted classificatory schemes, still have relatively important features in common.

In this way, the following relationship between the structural and the heuristic forms of evolutionary economics can be established. The difference between these two forms resides in the nature of the parallels that are being established between the relevant evolutionary biological and the relevant economic phenomena. If these parallels are based on the essential features used to type these phenomena, then it is a case of a structural evolutionary economics, and insights from evolutionary biology can be directly transferred to the economic case. If the parallels are based on features that fall short of allowing the phenomena to be identically typed, then it is (at best) a case of a heuristic evolutionary economics, and insights from evolutionary biology can only be used (at best) to suggest novel and plausible ideas to explore in economics. For these reasons, the structural form of evolutionary economics is epistemically stronger than the heuristic one. Since the structural form focuses on processes of the same type, whatever is known about this type of process in one context (evolutionary biology) also applies in the other (economics).[24] By contrast, since the heuristic form focuses on processes of different types, whatever is known about this type of process in one context (evolutionary biology) can merely suggest that something similar *may* be true in the other (economics) as well.

Note that none of this is meant to suggest that the existence of some peripheral similarities between an evolutionary biological and an economic phenomenon *must* make the consideration of the former heuristically useful for the investigation of the latter. Whether the consideration of any set of peripheral similarities is heuristically useful needs to be shown on a case-by-case basis. This depends on the number and importance of these peripheral features—issues that only a detailed analysis of the particular issues at stake can reveal. The point is just that consideration of such peripheral features *might* be heuristically useful—though falling short of establishing a structural parallelism between the two cases.

4. The Importance of the Three Forms of Evolutionary Economics

Putting all of this together, three points can be made. First, there is no question that there are important differences between the three forms of evolutionary economics. They have different aims and methods, and should thus be seen to be different in nature. In the first instance, therefore, they need to be discussed and assessed separately from each other. This is important, as they are sometimes conflated, and sometimes arguments switch back and forth between them without marking these switches clearly. In turn, this can make it difficult to determine what kind of conclusion the relevant arguments are trying to derive, and how compelling these conclusions are. A good example is the following quote from R. Nelson et al. (2018, p. 8):

> [Evolutionary economics] sees the configuration of economic activity at any time as the current result of an evolutionary process whose workings over time have generated a variety of different behaviors which vary in effectiveness, which have been winnowed but not completely (among other things because of the continuing innovation going on). Evolutionary economists believe that this orientation provides a much better basis for understanding how modern capitalist economies work.

In the beginning of the paragraph, it is suggested that economic activity is a product of an evolutionary process. In the middle, it is suggested that economic activity constitutes the basis for an evolutionary process in its own right. At the end, it is suggested that an evolutionary perspective provides a basis for understanding something about how modern capitalist economies work. However, these are three separate issues: economic activity may be a product of an evolutionary process without itself being such a process, and by using evolutionary biology as a heuristic device, we may be able to understand better "how modern capitalist economies work"—though without it being the case that economic activity is either a product of an evolutionary process or type-identical with such a process. Hence, it is best to keep these three argumentative paths separate, at least initially.[25] In short, paying attention to the differences between the three forms of evolutionary economics is very important in order to be clear both about what kinds of conclusions the relevant arguments yield, and how strong these conclusions are.

However—and this is the second point to be made here—it is also important to keep in mind that that there are deep connections between the three forms of evolutionary economics. In particular: (i) both the structural and the heuristic forms rest on similarities between evolutionary biology and economics that are being appealed to—they just differ over the nature of these similarities; (ii) both the structural and the evidential forms provide reasons for thinking that a given economic hypothesis is true—it is just that the structural form provides (generally) stronger reasons, but is also based on a tighter set of linkages between the two subjects; and (iii) both the evidential and the heuristic forms can advance discussions in economics by focusing our attention on specific economic hypotheses—they just differ over whether they provide a reason for thinking that these hypotheses are true, rather than for suggesting novel hypotheses that *may* be true. Given these interrelationships among the different forms of evolutionary economics, a comprehensive analysis of evolutionary economics is well advised to cover all of them; their strengths and weaknesses come out especially clearly by contrasting them with each other.

The third point to note here is that together these three forms of evolutionary economics cover all the major ways in which the two subjects can be linked. In order to combine evolutionary biological and economic thinking, we need to see these two as (1) concerned with what are (ultimately) the same processes; (2) informationally linked, so that knowing something about one of them tells us something about the other (even if this falls short of them being concerned with the same processes); or (3) sharing similarities that allow us to generate novel and plausible hypotheses about one of them by considering the other.[26] These three ways largely carve up logical space here. It is not clear what other ways of connecting these two subjects there could be.[27]

It is thus unsurprising that most of the existing work in evolutionary economics can be seen to make use of (1)–(3) in one way or another. So, for example, Nelson and Winter's (1982) argument that the spread of innovative productive techniques through large firms is (often) an evolutionary process is an instance of (1). By contrast, the argument of D. S. Wilson, Ostrom, et al. (2013, p. S30) that the principles identified by Ostrom (1990) for enabling the efficient management of common-pool resources follow from "the evolutionary dynamics of cooperation in all species and the biocultural evolution of our own species" is an instance of (2). The idea is that these principles are products of certain kinds of evolutionary processes (the study of which thus tell us more about these principles). Similarly, Lo's (2017) argument that, in order to understand the investment decisions that humans make, it is useful to see humans as having evolved so as to make decisions dynamically and by reliance on simple heuristics, falls into this category (see also V. Smith, 2007). Lastly, Nelson et al.'s (2018) claim that in order to provide "good guidance to thinking and research regarding all aspects of the economy" (p. 228) "a broad evolutionary theory is likely to be much more fruitful" (p. 219) is an instance of (3).[28]

Of course, this brief sketch of existing work in evolutionary economics does not amount to a comprehensive literature review. However, as noted in the introduction, this is not the purpose of this book either. Rather, the point made here is just that the division of evolutionary economics into a structural, an evidential, and a heuristic form covers the spectrum of ways of bringing evolutionary biological considerations to bear on economic debates. Nothing major is left out by focusing on this three-part division of evolutionary biologically grounded economics. Given this, what is needed next is a closer look at how these three different forms of evolutionary economics work in practice. What promises and challenges arise when we *actually* try to apply evolutionary biological thinking in economics in one of these three ways? This is what the rest of the book is aiming to do. As noted in the introduction, to do this, the book focuses on applications that (a) are particularly conducive to bringing out the methodological lessons to be derived; (b) exhibit the breadth of economic issues that can be analyzed from an evolutionary point of view; and (c) have not been assessed in detail to date. However, it is important to emphasize again that this does not mean that there are not many other examples that could not also be discussed here. The point is just that, in order to be manageable, this book focuses on five specific examples with features (a)–(c).

VI. Conclusions

Evolutionary economics divides into (at least) three different—but connected—forms: a structural one (which explores the idea that the two subjects investigate the same type of processes), an evidential one (which explores the idea that human economic decision-making can be better understood by investigating how it has evolved), and a heuristic one (which explores the possibility of using evolutionary biology as a source of suggestions for how to investigate economic issues). These forms increase in their epistemic degree of strength from the last to the first—the heuristic form only provides novel and plausible hypotheses to consider, the evidential form provides evidence, and the structural form allows for the direct application of evolutionary biological insights in economics. However, all three can, at least potentially, play an important part in advancing discussions in economics.

Noting that there are these three forms of evolutionary economics is not just inherently interesting, but it is also methodologically important. Since these three forms have different aims and methods, a critical assessment of one need not translate into a critical assessment of another. In turn, this implies that these three different forms, at least to some extent, need to be investigated separately from each other. That said, the fact that there are also important connections between these three forms suggests that a comprehensive and clear analysis of the field need to consider all three of them.

Finally, all three forms of evolutionary economics are at least inherently coherent. However, to what extent they are *in fact* successful in achieving their aims still needs to be seen. As is so often the case, the proof of the pudding is in the eating—only a detailed look at instances of the three forms can tell us whether they are in fact plausible (see also Wimsatt, 2007, p. 26). This is what the rest of book seeks to achieve.

Suggested Further Reading

For a contemporary overview of a narrow form of evolutionary economics, good places to start are R. Nelson et al. (2018), Hodgson and Knudsen (2010), Hodgson (2007b, 2019). Another good overview is the 2013 special issue (90S) of *The Journal of Economic Behavior & Organization* that features a wide-ranging set of discussions of some of the structural, evidential, and heuristic ways evolutionary biology and economics can be linked. Relatedly, the website of David Sloan Wilson's Evolution Institute (https://evolution-institute.org/projects/evonomics/) contains a number of useful resources for further reading on the intersection between evolutionary biology and economics. Chao et al. (2013) contains some interesting work on the role of mechanisms and causality in evolutionary biology and economics. Ofek (2013) is an evidential evolutionary economic take on the explanation of the human disposition to trade. Güth and Kliemt (1998) is a good introduction to the "indirect evolutionary approach" that looks towards evolutionary biological modeling to understand the nature of economic preference formation.

Notes

1. While the connection between these two subjects is, at least in principle, a two-way affair—the goal could be an advance in our understanding in either one subject (or indeed both of them)—the focus is typically on using evolutionary biology to push forward debates in economics (for an exception, see e.g. Ofek, 2001). This focus is being maintained in the present book. Note also that economists could of course draw on other subjects (such as physics) to advance debates in economics as well (see e.g. Richmond et al., 2013). However, the focus here will be on combinations of evolutionary biology and economic (this is not to say, though, that some of what follows does not have implications for "econophysics" or the like as well).

2. Of course, Darwin even famously saw connections between evolutionary biology and economics—though he did not try to use these connections to advance economics.

3. Another perspective that differs from the one in this book is that of Chao et al. (2013). The latter is more narrow, in that it focuses only on issues surrounding mechanisms and causality in biology and economics, but it is wider, in that it considers anything to do with causality and mechanism—even in the fields individually, and even if it is not specifically about evolution.

4. A quick point of terminology: some writers may prefer to restrict the label "evolutionary economics" to the narrow form used here (see e.g. Witt, 2008). Yet other authors (e.g. D. S. Wilson) have coined the label "evonomics" for a slightly different, but also substantively restricted, form. Both of these sets of authors may prefer a different label for the broader project at the forefront of this book—maybe something like "evolutionary biologically grounded economics." For ease of exposition, though, I shall stick "evolutionary economics" as the name of the general project of bringing evolutionary biological considerations to bear on economic debates. However, this is purely a verbal point, and all the conclusions of the book can easily be reformulated using a label like "evolutionary biologically grounded economics."

5. Relatedly, there is no assumption here that all aspects of evolutionary biology (including, e.g., epigenetics or evo-devo: see e.g. Wagner & Altenberg, 1996; Jablonka & Lamb, 2005; Stotz, 2006) need to be able to be connected to some part of economics. The issue is just whether and how *some* evolutionary biological insights can be put to use in economics.

6. This form of evolutionary economics could thus also be seen as a "reductivist" or "unificationist" project. However, since the characterization of such projects is controversial (Churchland, 1985; Sarkar, 2005; Kitcher, 1989; Wimsatt, 2007), and since further discussion of this characterization is not relevant for the rest of this chapter (or indeed the book), I do not consider this further here. A closely related idea also underlies what Hausman (1989) calls "arbitrage arguments."

7. There are some important questions about the nature of and differences between replication and reproduction, but for present purposes, these questions do not need to be addressed. Much the same goes for the notion of "population." (For more on both of these issues, see e.g. Godfrey-Smith, 2009.)

8. This is also underwritten by the fact that it is not entirely clear exactly which entities are "biological" entities. Viruses can evolve—though they are not clearly living things—and some have argued that species (Vrba, 1984), cells (Pradeu, 2012), ecosystems (Cropp & Gabric, 2002), or even crystals (Maynard Smith & Szethmary, 1995) can evolve—none of which are straightforwardly "living things" either. See also Godfrey-Smith (2009).

9. In the background of this answer is a broadly realist view of science (see e.g. Worrall, 1989; Chakravartty, 2017). If it is questioned whether there are *any* privileged ways of typing phenomena—so that all such typing is in the eyes of (communities of) individual researchers only—then this answer of how to determine whether an application of evolutionary biological thinking in economics is part of the structural form will not have a foot to stand on. However, this commitment to some kind of scientific realism is in keeping with much of (philosophy of) science, and thus not something that creates a major problem for the present project of investigating the promises and challenges of evolutionary economics. At any rate, further discussion of scientific realism will be left for another occasion.

10. It is of course possible to use the term "evolution" in many different ways, including in ways that differ from (i)–(iii) (see e.g. Witt, 2003, 2008). However, the point in the text is just that *in evolutionary biology* there is

widespread agreement on evolution being a set of population-level processes surrounding reproduction (see e.g. Futuyma, 2009; Godfrey-Smith, 2009; Sober, 2000, 1980).

11. Hodgson and Knudsen (2010, pp. 94–104) use the labels "successor selection" and "subset selection" (derived from Price) for this distinction.

12. Hodgson and Knudsen (2010) argue in favor of the need for replication. However, since they employ a very broad sense of replication, this is a more of semantic than a substantive difference to the point made in the text. For reasons why the replicator view—on a narrower reading—is not a compelling general account of evolutionary processes, see e.g. Godfrey-Smith (2009) and Sober (2000, 1984).

13. Of course, any uncertainties about the relevant evolutionary biological hypotheses are being maintained in the transfer to the economic case. It is just that there are no *further* uncertainties that are being introduced during this transfer. See also section V which follows.

14. This point can be generalized to some of the work employing evolutionary and/or rational game theory (Alexander, 2009). However, it is important to note that the latter kind of work is not exclusively tied to the structural form of evolutionary economics, and that it can be used to underwrite all three forms of evolutionary economics. See also note 28 that follows.

15. Note that it may also be thought that it is interesting to point out that the two subjects study the same processes, without there needing to be a transferal of insights from one to the other. For example, this may allow for a *unification* of the two subject matters (see e.g. Hammerstein & Noe, 2016; Driscoll, 2018). While perfectly reasonable as such, the epistemic value of such purely unificatory projects is limited. Without there being a transferal of insights, the mere fact of unification is not of major epistemic interest (though see also Kitcher, 1989). At any rate, while I *focus* on the stronger, transferal-based view in the rest of this book, most of the conclusions also apply to the weaker, purely unifying view.

16. Here and in what follows, I do not use the term "mental mechanism" in the technical sense used e.g. by Machamer et al. (2000) or Bechtel and Abrahamsen (2005), but rather as a stylistic variant of "psychological structures" or the like.

17. Also important is that this point is true independently of the debate surrounding how "microfoundationalist" the subject is required to be—i.e. whether all of economics should be derived "from the ground up" by starting with individual decision making (for more on this, see e.g. Hoover, 2010, 2015, 2001, 2009; Elster, 1982; Hodgson, 2007a). Even strongly microfoundationalist projects can and must go beyond individual decision making to understand various economic phenomena, such as the determinants of economic growth. Microfoundationalist projects just insist that we need to *start* with individual economic decision making, not that we need to *end* with it (see e.g. Romer, 1990).

18. This can involve mathematical modeling of one kind or another. For an intriguing recent example of this kind of work, see e.g. Heller and Mohlin (2019). They argue that one factor driving the evolution of increased mindreading abilities is the possibility to deceive those with lesser mindreading abilities about their preferences in strategic interactions. Of course, for such mathematical results to tell us something about how humans actually make decisions in strategic interactions, the assumptions of the modelling framework must be connected to the conditions prevailing in human evolutionary history (as also made clear by Heller & Mohlin, 2019), thus (re-)introducing the issues laid out in the main text. For more on evolutionary modeling about

human preferences, see also Gale et al. (1995), Binmore (2007), L. Samuelson and Swinkels (2006), L. Samuelson (2002, 2001), A. J. Robson (2001b), A. J. Robson (2003), A. J. Robson (2002, 2001a), and A. J. Robson and Samuelson (2008). For another major version of this kind model-based approach towards evidential forms of evolutionary economics, see the "indirect evolutionary approach" towards the determination of preferences (see e.g. Güth & Kliemt, 1998).

19. In this and several other regards, the present form shares some communalities with work in evolutionary psychology: see also Schulz (2011b, 2018b).
20. For a somewhat parallel proposal about the methodology of evolutionary psychology, see e.g. Machery (forthcoming).
21. Brandon's (1990) discussion of "how possible" explanations makes some very similar points. Wimsatt (2007, pp. 76–78) also notes that heuristics are not closely truth-linked, and often *transform* problems.
22. Indeed, one could also look towards evolutionary biology to help explain why economists—as human scientists—have the particular biases they do in fact display. However, I will not discuss this further here. See Wimsatt (2007) for more on this.
23. This assumes a certain kind of "epistemic monotonicity" in evidential relations—the epistemic value of one piece of evidence is not being altered once other pieces become added to the picture. While this kind of monotonicity need not characterize all evidential relations, there is no reason to think that it does not apply to the cases of interest here.
24. As was true when it comes to the relationships between the evidential and the heuristic forms, it is of course also true that in a trivial sense the existence of structural parallelisms can be used to suggest novel and prima facie plausible hypotheses to consider further. The point in the text is just that it can do more than that too.
25. "At least initially," since it is possible to follow Okasha (2011), for example, and reason as follows: if it is fitness maximizing to make decisions in a way that is structurally parallel to a biologically selective, fitness maximizing process, then organisms that are under selection will be evolutionarily pushed to make decisions in a way that is structurally parallel to a biologically selective, fitness maximizing process. The point in the text is just that it is important to recognize that there are two separate steps involved in this argument: one concerning economic decision making being the product of a biological selection process, and one concerning economic decision making itself being a biological selection process.
26. Note also that (1) and (2) may allow for evolutionary biological *explanations* of economic phenomena—though this will depend on both the details of the case in question and a defense of particular account of explanation. For example, whether natural selection can be seen as a population-level causal process or mechanism (and thus as the basis for causal or mechanistic explanations)—rather than a statistical summary of individual level causal processes—is currently a point of contention in the literature (Millstein, 2013; Walsh, 2015; Ariew et al., 2014; Sober & Shapiro, 2007). Fortunately, this need not be resolved here. For present purposes, it is sufficient to focus on the existence of structural, evidential, or heuristic connections between evolutionary biology and economics. Whether and when these amount to explanatory relations *as well* can be left for another occasion.
27. The one form of evolutionary economics that is left out of this division is one that takes a "meta-level" view of science and social science, and sees the appeal to evolutionary theory as a new kind of "cosmology" or metaphysics that provides for a new, process-based way to tell economic stories (see e.g.

Gowdy et al., 2013; see also Herrmann-Pillath, 2013; Dopfer & Potts, 2004). However, this sort of view moves far from a science-based view of the issues here, and is best left for another occasion (see e.g. Herrmann-Pillath, 2013). At any rate, as noted in the text, most of the work linking evolutionary biology and economics falls under one of the three categories laid out here. Similarly, this division leaves out purely metaphorical or stylistic appeals to evolutionary theory in economics (as when it is said that we need to "to evolve a set of policies that increase the prevalence of wellbeing": Biglan & Cody, 2013, p. S158). However, in these cases, the appeal to "evolution" is not really meant to carry with it biological implications, and is just a stylistic variant for something like "choose" or "discover."

28. Note that work employing evolutionary game theoretic modeling to address economic questions can fall under any of (1)–(3). So, a number of the arguments of Chapters 2 and 3 are related to some structural versions of this work (see also Alexander, 2009; Gruene-Yanoff, 2011; R. Nelson & Winter, 1982); some of the arguments of Chapters 4 and 5 are related to other, more evidential versions of this work (see also Boyd & Richerson, 2005; Bicchieri, 2006; Skyrms, 2004); and finally, some of the arguments of Chapter 6 are related to more heuristic versions of this work (see also Galla & Farmer, 2013). See also the remarks on evolutionary modeling about preferences in note 18.

2 Economic Choice as a Selective Process (The Structural Project I)

I. Introduction

The attempt to connect evolutionary biology and economics using a posited parallelism between natural selection and economic decision-making is one of the most popular and extensively discussed parts of evolutionary economics (see e.g. Cooper, 1987; Stearns, 2000; Okasha, 2011; Gintis, 2009; see also Hagen et al., 2012; Alexander, 2009; A. J. Robson, 1996; Orr, 2007).[1] The idea underlying this way of bringing evolutionary biological considerations into economics is that the nature of some of the key processes studied in the two subjects seems (near) identical: in both evolutionary biology and economics, some of the core processes at stake are, on the one hand, optimizing in nature, and on the other, directed at (nearly) the same quantity (though this quantity happens to be labeled differently in evolutionary biology and economics) (Sterelny, 2012b; Okasha, 2011, 2007; Orr, 2007; Rosenberg, 2000, pp. 134–135; Cooper, 1987, 1989; Becker, 1976; Kenrick et al., 2009; Gintis, 2009; Stearns, 2000; A. J. Robson, 1996).[2] As noted in Chapter 1, if justifiable, the existence of this close parallelism would be important, as it would suggest that insights from one domain can be directly carried over into the other, thus opening up new avenues for progress in the two subjects. My goal in this chapter is to assess to what extent—i.e. under which circumstances—the postulation of a close natural selection/economic choice parallelism is plausible.

To do this, I proceed somewhat indirectly. After laying out the basics of the attempt to link evolutionary biology and economics through a natural selection/economic choice parallelism, I analyze the relationship between two phenomena: one evolutionary biological and one economic. On the one hand, there is the fact that organisms sometimes change their environments in order to make these environments better suited to what they happen to be, rather than change themselves to make themselves fit better to what their environments happen to be—a phenomenon that has become known as "niche construction" (Lewontin, 1982; Odling-Smee et al., 2003; Laland & Sterelny, 2006). On the other hand, there is the fact that agents sometimes change their preferences about what they want to achieve so that they come to match what the world happens to be, rather than doing something to the world so that it comes to match what their preferences happen to be—a phenomenon that has become known as involving "adaptive preferences" (Elster, 1983; Bovens, 1992; Sen, 1995; Nussbaum, 2000; Zimmerman, 2003; Bruckner, 2007; Hill, 2009).[3] The reason for analyzing these two phenomena is that they turn out to be useful stepping stones in the formulation of a general criterion of when the postulation of a natural selection/economic decision-making parallelism is defensible. (Moreover, in the course of this discussion, we can shed some new first-order light on the concepts of niche construction and adaptive preferences.)

The chapter is structured as follows. In section II, I make clearer what positing a natural selection/economic choice parallelism entails. In section III, I assess the extent to which fitness and preference values should be expected to be ordinally equivalent. In section IV, I begin to consider the question of whether fitness and preference values (independently of whether they are ordinally equivalent) function in the same way in evolutionary biology and economics by assessing the extent to which they are optimized in the same way in the two subjects. I deepen the analysis of this question by comparing the logic of niche construction and adaptive preferences in section V. In light of this discussion, in section VI, I bring the previous insights together by setting out the kind of work that can—and the kind of work that cannot—be done by connecting evolutionary biology and economics using the natural selection/economic choice parallelism. I conclude in section VII.

II. The Natural Selection/Economic Choice Parallelism: What It Is and Why It Matters

The (supposed) structural parallelism between selective evolutionary processes and economic decision-making can be broken down into two claims:

(a) In many cases of interest, fitness and preference values are ordinally equivalent.
(b) In many cases of interest, fitness and preference values function in the same way in evolutionary and economic processes.

These two claims will be spelled out in more detail in section III (for claim (a)) and sections IV and V (for claim (b)); for now, a brief explanation of their content is all that is needed.[4] Before giving this explanation, it is useful to note that there is debate over how to conceptualize an agent's preference values; for example, whether this is to be done using utilities, prospects, or regrets (Hausman, 2012; Angner, 2018, 2016). As will be made clearer in the sections that follow, and in Chapters 4 and 5, this is an important issue that needs to be addressed. However, for present purposes, a detailed treatment of this is not necessary. To make the exposition easier, therefore, I follow much of the rest of the literature in this context and largely focus on *utilities*. However, it is important to emphasize that this is done for expository purposes only, and should not be read as implying that economic decision-making *needs* to be seen to be based on expected utility maximization.

With this in mind, note that claim (a) says that, in many cases of interest, the fitness and utility values of a set of traits will depend on the same sorts of facts (e.g. the caloric values of different foodstuffs that can be obtained by different behavioral strategies), and will thus rank the traits in the same way. As Cooper (1987, p. 397) put it: "Darwinian fitness and decision-theoretic utility are [here] seen as intimately related and almost the same thing. Specifically, expected utilities are conceived to play the role of hypothetical organismic fitness estimates, or internal constructs that tend to track fitness."

Claim (b) says that the role that fitness values play in evolutionary biological processes is parallel to that of utility values in economic processes: in both cases, it involves a kind of constrained optimization, where the option that is evolutionary biologically or economically selected has the highest fitness/(expected) utility value among the available ones. As Okasha (2011, p. 84) puts it: "My strategy will be to exploit an analogy between utility in rational choice and fitness in Darwinian evolution. (The former is the quantity that the rational agents try to maximize; the latter, that natural selection tries to maximize.)"

In this way, the two sciences can seemingly be seen to deal with (virtually) identical processes: they study processes that are optimizing in nature, and which range over closely related (or even identical) quantities. If correct, this would be an important insight, as it would be a powerful way of making progress in our understanding of various economic questions. To see this, briefly consider an example of the kind of evolutionary economic work that can be done with the help of claims (a) and (b).

This example concerns the debate of how to understand the attitudes towards risk displayed by people: should economic decision makers be seen to be risk-neutral (i.e. indifferent to the variance associated with their decision outcomes), risk-averse (i.e. preferring choices with lower variance in outcomes, ceteris paribus), or risk-loving (i.e. preferring choices with higher variance in outcomes, ceteris paribus) (Mas-Colell et al., 1996; Okasha, 2007)?[5] Note that there is no question *that* people sometimes act in ways that are risk-averse (they take out insurance), sometimes in ways that are risk-loving (they gamble), and sometimes in a way that is risk-neutral (they seek out whatever investment vehicle maximizes their returns when it comes to saving for their retirement).[6] The real question concerns how these behaviors are to be explained. Here, there are three possible explanations that have been proposed.

First, people's attitudes towards risk could be seen to stem from the shape of the utility functions that underlie their choice behaviors (Friedman & Savage, 1948; de Finetti, 1979; Morgenstern, 1979). In turn, this is due to the fact that non-linear utility functions directly lead to decision-making that is risk-averse, risk-neutral, or risk-loving—depending on the nature of the non-linearity of the utility function. To see this, consider the following graph:

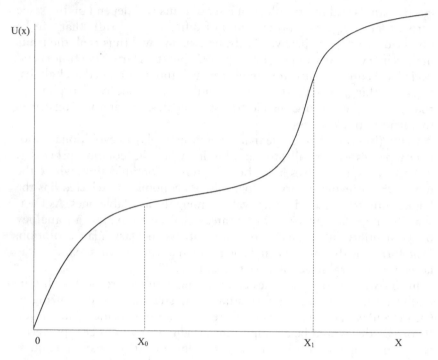

Figure 2.1 A Friedman-Savage utility function

In this graph, the following can be seen to hold. In between points $x = 0$ and $x = x_0$, it is the case that $E(u(X)) < u(E(X))$, where $E(.)$ is the expectation (arithmetic average), and X is any set of points in $[0, x_0]$. This means that the utility of the expected outcome of a gamble is higher than the expected utility of the gamble—people prefer getting $5 for sure than a 50% chance of winning $10 and a 50% chance of winning $0. In turn, this implies that people are risk-averse: they would decline a gamble with an expected value of $5 in favor of a certain option with a lower value. The exact reverse is true in between points x_0 and x_1: people are risk-loving. Finally, in between points x_1 and infinity, they are risk-averse again. Of course, many other shapes of the utility function are possible as well—including parts that are linear and thus risk-neutral. The key point is just that, on this picture, people's attitudes towards risk are seen to stem simply from the shape of their utility function.

The second way in which people's attitudes towards risk can be understood is by appeal to the nature of economic decision-making itself. In particular, it may be that economic decision-making is not to be seen to be based—purely or at all—on what maximizes expected utilities, but on something else. (I return to the details of these alternative theories in the sections that follow and in Chapters 4 and 5.) For example, Gigerenzer and his colleagues (see e.g. Gigerenzer & Selten, 2001; see also Simon, 1957) have argued that economic agents merely "satisfice"—they use simple decision rules that, in the right environments, lead to sufficiently good choices. If so, then risk aversion and risk love can result from the fact that economic decision-making, in its very nature, goes beyond the expected utilities of the options in question. This differs from the first case, in that we are here not (necessarily) committed to a specific nature of the underlying utility functions: the utility function can be entirely linear, and still economic decision-making can be risk-averse or risk-loving. In this second view, what underlies economic decision-making is not (just) the shape of this utility function, but something else (as well).[7]

The third and final way of understanding attitudes towards risk is by seeing them, wholly or partly, as the result of "choice-external" influences on economic decision-making: while economic decision-making is inherently risk-neutral (say), some economic behaviors are at least partly "non-chosen"—i.e. the outcome of mechanisms separate from economic choice. Exactly what these mechanisms are is controversial, but for present purposes, this can be left open. So, it may be that purely economic decision-making is entirely risk-neutral, but that people's emotional states can interfere with their ability to make economic decisions in such a way that they end up acting in a risk-averse or risk-loving manner (see e.g. Rosenberg, 2012, chap. 3; R. Frank, 1988). This differs from the previous two accounts, as, in this picture, risk aversion or risk loving should not be seen as the result of the processes underlying economic decision-making *themselves*, but as the result of something else *interfering* with these processes.

Exactly how to theoretically capture human attitudes towards risk—i.e. when which of the three options is most plausible—is still an open question (Okasha, 2007; Schulz, 2008). For present purposes, what is most important about this ongoing debate is that an evolutionary economic perspective based on claims (a) and (b), as stated earlier in this section, could be used to make some progress in resolving it.

To see this, begin by noting that evolutionary biologists have found that, in general, the *variance* of a reproductive or survival strategy is a crucial factor determining its fitness. While there are environments in which fitness of an organism is just equal to the expected number of its offspring (perhaps modulated so as to accommodate kinship relations:

Gardner et al., 2011), in many environments, organisms will be higher in fitness the *lower* the variance is of their disposition to produce offspring (Gillespie, 1977; Sober, 2001; Pence & Ramsey, 2013a; McNamara, 1995; A. J. Robson, 1996; Curry, 2001). This is important, for if claims (a) and (b) are accepted, this fact about the variance/fitness connection has some direct implications for how to understand attitudes towards risk in economics.

In particular, claim (a) and the variance/fitness connection together imply that utility (or more generally preference) functions should generally be seen to be concave (with linearity a limiting case of this: Orr, 2009). Claim (b) implies that economic decision-making is *not* based on something other than utility (or more generally preference) values. Putting this together, this leads to the conclusion that economic agency should be seen to be biased towards risk aversion: economic agency should be seen to be based on utility (or more generally preference) maximization and utility (or more generally preference) functions and, by and large, should be seen to be concave (Orr, 2007; Okasha, 2011, 2007; Schulz, 2008). This further implies that any risk aversion that cannot be accommodated in this way, as well as all other attitudes towards risk, need to be seen to be the result of choice-external processes that interfere with the processes underlying economic decision-making per se (see also Schulz, 2008). This thus confirms a combination of options one and three for accounting for human attitudes towards risk: economic decision-making should be expected to be risk-averse; departures from this inherent risk aversion are due to other biological, cognitive, social, or other processes interfering with the mechanisms underlying economic decision-making. Put differently, we should expect economic decision-making itself to be risk-averse (or at best risk-neutral). All other attitudes towards risk are the result of something interfering with pure economic decision-making.[8]

As will be made clearer in the following and in Chapters 4 and 5, there is much else that needs to be said about these issues. However, what matters for present purposes is just that this way of resolving the debate surrounding attitudes towards risk in economics illuminates the potential value of the appeal to a natural selection/economic decision-making parallelism. Of course, whether this value can in fact be realized depends entirely on whether the elements making up the natural selection/economic decision-making parallelism—i.e. claims (a) and (b), as described previously—can in fact be maintained. Investigating this further is the aim of the rest of this chapter. To do this, consider these two claims in more detail.

III. The Ordinal Equivalence of Fitness and Preference

Why should we think that, in many cases of interest, fitness and preference values are ordinally equivalent—i.e. that claim (a) is true? The core argument for this claim can be spelled out as follows (see Millikan, 1984; Papineau, 1987; for hints of a similar argument).

An organism's preferences represent how it evaluates the choice situation it faces (Hausman, 2012; but see also Angner, 2018). For this reason, the content of the organism's preferences is a key element of its decision process: to the extent that the organism acts on its preferences at all—and not on other mental states like reflexes (Schulz, 2018b)—this content determines what the organism will do. Given this connection between what an agent prefers and what it does, agents with preferences that match their biological needs—such as preferences for food, protection, and sex—are more likely to fulfil these needs than agents whose preferences do not match their biological needs. Put differently, agents that *prefer* to do things that are biologically good for them are more likely to in fact *do* things that are biologically good for them. This matters, as, for obvious (and nearly trivial) reasons, there is selection on organisms doing what is biologically good for them. Putting all of this together implies that it is reasonable to think that there is selection for organisms to have preferences for doing things that are biologically advantageous— for these preferences are a key source in getting them to act in ways that are biologically advantageous (Tooby & Cosmides, 1992; Buller, 2005, chap. 2).

Furthermore, it is plausible that an agent's preferences are heritable and thus could evolve by natural selection. While there is no reason to think that *every* preference of *every* agent is heritable, there are two reasons why it is plausible that *many* preferences of *many* agents are heritable. On the one hand, an agent's preferences are likely to shape the environment her offspring grows up in. People with a preference for peas over carrots are likely to give more peas than carrots to their children, making it more likely that the children will also favor peas to carrots (Sterelny, 2012a; Odling-Smee et al., 2003). On the other hand, at least some preferences have in fact been shown to be heritable in human and non-human animals (D. M. Buss, 2014; D. M. Buss & Schmitt, 1993). While neither of these reasons implies that *all* preferences are *always strongly* heritable, they do at least underwrite the claim that a sufficient interesting subset of them are at least somewhat heritable.

Putting all of this together implies that agents with preferences for engaging in biologically advantageous actions will be more likely to survive and reproduce than agents with different preferences—and their offspring are likely to share their preferences (see also Collins et al., 2016). Over time, we would thus expect that preferences that match biological needs will likely spread in a population of organisms. In this way, we obtain a reason for thinking that claim (a) is true.[9]

However, at this point, two objections to this argument might come to mind. First, it may appear that the reasoning just sketched is unconvincing, as it is based on a highly simplistic model of cognitive evolution.

As the research on gene-culture coevolutionary theory has shown, for a cultural species like our own, preferring what is popular or entertaining (though possibly maladaptive) can, at least at times, also be expected to spread in the population (Boyd & Richerson, 2005, 1985; Sterelny, 2012a; Boyd et al., 2011; Henrich, 2015; Henrich & McElreath, 2011, 2007; see also Chap. 5). So, it is reasonable to think that humans have not only evolved minds focused on learning from others, but minds that are focused on learning from *specific types of others*—such as those with high social status. In turn, this is plausibly due the fact copying those that have high social status makes it more likely that one copies traits that causally contribute to achieving high social status (Boyd et al., 2011; Henrich, 2015; Henrich & McElreath, 2011, 2007; Boyd & Richerson, 2005). If this is so, though, then traits associated with high social status can spread through a population—even if they have a neutral or detrimental effect on individual fitness. A classic example of this phenomenon is celibacy; if priests are high-social status individuals, and if being a priest is associated with being celibate, then celibacy can be maintained in a population, even though it is biologically disastrous for the individuals involved (Boyd & Richerson, 2005, 1985; Boyd et al., 2011; Henrich, 2015; Henrich & McElreath, 2007). For this reason, what an organism wants and what is biologically good for it should not always be thought to be identical. It is possible that preferences are culturally acquired, and therefore possibly non-adaptive (Güth & Kliemt, 1998; Chudek et al., 2013, pp. 436–437; Sterelny, 2012b).[10]

However, it turns out that these facts do not, in the end, fully invalidate claim (a)—though they do make clear that claim (a) is restricted in important ways. The reason for this is that it is also a key insight of gene-culture coevolutionary theory that the ability for social learning—and thus, for the acquisition of maladaptive preferences—can only be understood on the basis of the fact that it generally leads to adaptive behavior (Boyd & Richerson, 2005; Henrich, 2015; Schulz, 2018b). If most or all of the behaviorally most influential preferences, beliefs, or behaviors acquired from others were maladaptive, it would not be adaptive to learn from others at all. Social learning is itself a trait whose evolution needs to be explained.[11] Because of this, it is reasonable to assume that many preferences—even if they are socially acquired—do lead to adaptive behavior. Maladaptive preferences, at least very plausibly, exist and need to be taken into consideration when attempting to link evolutionary biology and economics through a natural selection/economic choice parallelism; however, they should not be seen to completely undermine the plausibility of claim (a). That claim was formulated with the caveat that "*in many cases of interest*, fitness and preference values are ordinally equivalent." The prior objection makes clear that this caveat needs to be taken seriously; claim (a) cannot be taken to be universally true. However, the objection does not imply that this claim has no feet to stand on whatsoever. With this in mind, consider now the second—and somewhat related—objection to claim (a).

This second objection concerns the fact that it appears easy to come up with preferences that seem to have no adaptive value whatsoever. So for example, a preference for buying or looking at O'Keefe over Klimt paintings (or the reverse) is not plausibly seen as adaptive; these kinds of behaviors seem to be biologically completely neutral. Much the same can be said about a vast number of other preferences: Doctor Who vs. Sherlock, house cats vs. domestic dogs, Cubs vs. White Sox. Again, therefore, the *general* ordinal equivalence of fitness and utility appears implausible.

However, this objection is also less of a problem than it might at first appear. While it may not be plausible to see a preference for O'Keefe over Klimt (say) as adaptive, it is plausible to see as adaptive preferences for (i) seeing beautiful things over non-beautiful things (though see Pinker, 1997; Cross, 2007, for a debate on this point); (ii) signaling social status, education, or wealth (D. M. Buss & Schmitt, 1993; Barkow, 1992); and (iii) attracting mates, shelter, or protection (D. M. Buss & Schmitt, 1993; Gangestad & Simpson, 2000). Importantly, it may well be reasonable to see a preference for O'Keefe over Klimt—like many other preferences (such as for high fashion or luxury goods)—as *an instance of* these latter preferences (i)–(iii).[12] Put differently, while many preferences may not at first seem adaptive, further analysis of them might reveal them to be adaptive after all.[13]

Putting these two responses together, what this implies is that the real question to be asked concerning claim (a) is: *for the preferences at stake in the inquiry of interest*, is claim (a) reasonable? Given this, the answer to this will sometimes be yes and sometimes be no—it depends on the particular preferences in question. While perhaps disappointingly even-handed, this conclusion is in fact substantial. On the one hand, it makes clear that the present structural form of evolutionary economics *can*, at least in these regards, get off the ground. There is no principled reason for thinking that connecting fitness and preference values can never be done—i.e. for thinking that claim (a) is always false. In fact, this suggests that it *may be quite often* tenable. On the other hand, though, the discussion here also makes clear that the caveat built into claim (a) needs to be taken seriously. The ordinal equivalence of fitness and preference values needs to be established on a case-by-case basis and cannot simply be taken for granted (see also Chudek et al., 2013, pp. 436–437; Sterelny, 2012b; Boyd & Richerson, 2005).

Therefore, it thus becomes clear that the reasoning underlying claim (a) is sufficiently cogent to suggest that some correlation between fitness and preference values will exist in at least some important domains of behavior. In turn, this is enough to make it interesting to consider what sort of connections between evolutionary biology and economics can be sustained with claim (a) in the background. Consider therefore the situation with regards to claim (b).

IV. The Roles of Fitness and Preference Values I: Local vs. Global Maximization

Claim (b) goes beyond claim (a) in stating that it is not just that fitness and preference values lead to equivalent rankings of a set of traits, but also that these rankings have the same sorts of consequences in the two realms. Specifically, the same sort of constrained maximization analysis that is at the heart of much evolutionary biology is said to also be at heart of much of economics (Grafen, 1999; Hammerstein & Hagen, 2005; Okasha, 2011). In more detail, the reasoning behind this claim can be spelled out as follows.

On the one hand, at any given time, natural selection can be said to operate by selecting that trait F_i, out of the set of currently available traits F, that has the highest net fitness.[14] In other words, it solves the equation

(NS) $\text{Max } [w(F_i) \mid F_i \in F]$

where $w(F_i)$ is the net fitness of having trait F_i. (Note that this is different from the claim that natural selection maximizes the average fitness in the population. The latter claim is widely known to be true only in special cases—e.g. in the absence of frequency dependent selection or altruistic traits. What (NS) claims is merely that, *at any given time*, the fittest available trait is selected; which trait that is can change over time. Note also that (NS) should be seen to concern *net* fitness. There might be two, even opposing, effects on fitness—e.g. one via the individual level and one via the group level. Here, though, only the net effect matters; for more on this, see Okasha, 2006.)

On the other hand, economic agency can be said to operate by choosing that action A_i, out of the set of currently available actions A, that has the highest preference value. Keeping to the expository focus on *utilities* as the favored conceptualization of preference values—a point to which I return momentarily—this implies that economic agency can also be said to operate by choosing that action A_i—out of the set of currently available actions A—that has the highest net (expected) utility. In other words, it solves the equation

(EA) $\text{Max } [u(A_i) \mid A_i \in A]$

where $u(A_i)$ is the net (expected) utility of doing action A_i. Slightly more specifically, according to the classical theories of choice used in economics, it is assumed that (i) an agent has a set of preferences over what the world should be like (and which comprise both cognitive and conative components—see Hausman, 2012); (ii) these preferences satisfy a number of axioms (such as transitivity, completeness, etc.); and (iii) she acts on the basis of these preferences (Hausman, 2012). According to the representation theorems of these theories of choice, it is then the case that the agent's actions can be seen as those that maximize her (expected) utilities (Savage, 1954; Luce & Raiffa, 1957; Jeffrey, 1983; Joyce, 1999).

As will be made clearer in Chapter 4 (and as was also hinted at in Chapter 1), there are three main ways of understanding the claims of economic theories of choice (whether classical or not) (Hausman, 2012; Schulz, 2011a; Ross, 2005). They are: (1) we could see the theories as providing a (possibly idealized) description of the agent's actual decision-making processes, (2) we could see them as predictive tools for an agent's choices without descriptive intent as far as her actual psychology is concerned, and (3) we could see them as providing a standard of rationality that choices or decisions have to satisfy. While distinguishing these three readings is generally very important, for present purposes, it is not necessary to single out a particular one of these readings. Hence, in what follows, it is left open how (EA) is to be interpreted.

In this way, the (alleged) parallelism between the way fitness and preference values function in their respective theoretical contexts becomes very clear. In particular, as long as claim (a) holds, (NS) and (EA) seem to be descriptions of exactly the same kind of process, differing merely in the *name* of the key variable involved. However, it turns out that there are a number of objections to claim (b)—and the reasoning leading up to it—that need to be considered before this claim can be accepted.

The first objection returns to a point that was hinted at earlier: it concerns the worry that economic agency should not be seen to be based on expected utility maximization after all. So, some authors have argued that economic agency is not based on the maximization of *expected utilities*, but on the maximization of something else instead—such as *prospects* or *rejoicings* (see e.g. Kahneman & Tversky, 1979; Loomes & Sugden, 1982; Angner, 2018, 2016). However, this point turns out to not be so important here: parallel arguments to the ones stated could be given that focus on other conceptualizations of preference values (like prospects or rejoicings) (Okasha, 2011; McNamara, 1995; A. J. Robson, 1996; A. I. Houston et al., 2007; Sinn, 2003). What matters for claim (b) is that economic agency is based on preference maximization; for this, it is not so important how this notion of preference maximization is spelled out (Angner, 2018; Hausman, 2012). Hence the possibility that economic agency should be spelled out in prospect or regret theoretic terms is not greatly problematic for the present argument. (I return to issues related to this point in Chapter 4; see also note 8.)

The fact that some authors have argued that economic agency is not about *maximization* at all, but about something else instead—such as satisficing—is more important (see e.g. Gigerenzer et al., 2000; Simon,

1957).[15] That is, several authors have argued that, however preference values are to be conceptualized, economic decision-making should not be seen to be based on maximizing these preference values, but on using these values in some other way to determine what the organism is to do. If so, then claim (b) obviously fails—and with it the natural selection/economic choice parallelism.

In response to this worry, it needs to be noted that it is indeed true that preference maximization is not always the most reasonable view of economic agency (Angner, 2018, 2016). This is a key issue that is at the heart of Chapter 4 (and to some extent, Chapter 5), and will thus be discussed in more detail there. However, at this point, all that matters is that it is *also* true that preference maximization *sometimes does* appear to be the most reasonable view of economic agency. While seeing preference maximization as the only foundation of economic agency is not plausible, neither is it plausible to discard it completely. In particular—as will also be made clearer in Chapter 4—it is at least reasonable to see preference maximization as describing an *aspect* of economic decision-making.

In turn, this implies that this first objection to claim (b) fails to completely invalidate this claim. Rather, it merely points to a restriction in its scope: fitness and preference values function in the same way in evolutionary and economic processes only to the extent to which it is plausible to see economic decision-making as based on the maximization of preferences in the first place. While this restriction is unlikely to always be satisfied, it is likely to sometimes be satisfied. Hence, as was the case with claim (a), this restriction in the scope of the natural selection/economic decision-making parallelism, while not negligible, does not render the latter irrelevant.

This brings us to the second objection to claim (b). This objection concerns the fact that, even if it is accepted that both (NS) and (EA) are optimizing processes, they are optimizing in a different way (Sober, 1998). Specifically, evolution by natural selection is based on a "hill-climbing" procedure—the optimal trait is found by successive, small improvements to an existing trait, whereas economic agency as laid out in (EA) is based on a globally optimizing procedure—the optimal action is found by determining *whichever* action has the highest preference value. This matters, as it implies that evolution by natural selection need not reach the globally fittest trait; local fitness maximization is all it can directly accomplish. (It will reach the global optimum with probability approaching 1 only as the population size reaches infinity.)

However, by itself, this, too, is not a sufficient reason to give up on claim (b) altogether. In the first place, it is not entirely clear that economic agency needs to *always* be seen to be based on global optimization. It is plausible that economic decision-making will at least sometimes be based on local optimization only (see e.g. Busemeyer & Townsend, 1992, 1993). So, it is plausible that, at least at times, economic agents should only be assumed to be searching through the most obvious set of alternatives for the one that maximizes their preferences, rather than through *all* of the available options (see also Kahneman et al., 1982; Gigerenzer & Selten, 2001). Given the cognitive limitations of actual humans, this kind of restriction appears very reasonable. Second, the difference between local and global optimization disappears if the relevant fitness functions/ preference values are monotonically increasing, or if the population size is sufficiently large to make the likelihood of missing the global maximum negligible (Sober, 1998, 2008, chap. 3; Grafen, 1999). Of course, neither of these conditions is universally applicable, but collectively, they do cover a reasonably large set of cases. For this reason, this disanalogy in how fitness and preference values function may well be much less widespread than it might at first appear.

Altogether, this second objection also fails to fully invalidate claim (b). However, it adds further restrictions to the scope of this claim: in particular, this objection shows that this claim—and thus the natural selection/ economic choice parallelism it helps to set up—is limited to cases where economic agency is only based on local maximization of preference values, or where natural selection is based on monotonically increasing fitness functions. While these conditions are certainly stringent, there is no reason to think that they are not at least sometimes satisfied.

Therefore, while it turns out that there are a number of significant restrictions of the scope of claim (b), these restrictions do not amount to a reason to give up on this claim completely. At least in a number of important circumstances, it is not obviously false to think that, apart from the change in the name of the variables, fitness and preference values function in exactly the same way in evolutionary biology and economics. However, as I show in the next section, there is still more that needs to be said about this claim. Even if economic agency can reasonably be seen to be based on preference maximization, and even if local and global optimization lead to the same outcome in those case, there are still more concerns with claim (b) that need to be addressed.[16]

V. The Roles of Fitness and Preference Values II: Niche Construction vs. Adaptive Preferences

To bring out the further concerns with claim (b), I want to suggest it is useful to consider the relationship among two independently interesting phenomena in evolutionary theory and economics: niche construction and adaptive preferences. The reason why it is useful to compare these two phenomena is that, given claims (a) and (b), they seem to be structurally similar non-standard applications of (NS) and (EA), respectively.[17] In particular, both niche construction and adaptive preferences seem to be cases where the initially given fitness/preference values of the options (traits or actions) are being changed. Instead of an organism maximizing its fitness/preferences by adopting the option with the highest such value, the values of these options are changed such that the option the organism happens to have adopted ends up having the highest value. Importantly, though, while the two phenomena thus seem to be structurally related, their evaluation is quite different in the two sciences. In particular, most evolutionary biologists and philosophers see nothing problematic in principle about cases of niche construction, whereas most economists and philosophers find cases of adaptive preferences to be theoretically dubious. It is for this reason that comparing these two cases is so interesting: they appear to be structurally similar, but their evaluations are very different.

1. Niche Construction and Fitness Maximization

For a long time, the typical dynamics of evolution by natural selection have been conceived of as follows (Lewontin, 1982, 1983; Brandon, 1990, pp. 6–9; Godfrey-Smith, 2009): the environment is structured in such a way as to pose some "adaptive problem" to some population of organisms; there is variation in the population with regards to responses to this adaptive problem; finally, to the extent that these responses are heritable, the most adaptive among them gets selected—i.e. spreads in the population.

However, recently, a number of researchers have called for an emendation and revision of this picture, and believe a greater emphasis should be placed on organisms changing their environments (Lewontin, 1982, 2000; Brandon, 1990, pp. 68–69; Godfrey-Smith, 1996; Laland & Sterelny, 2006; Sterelny, 2003, 2012a; Odling-Smee et al., 2003; Odling-Smee, 1988). In particular, instead of only seeing populations of organisms as changing themselves in the face of an adaptive problem, they should also frequently be seen as *dissolving* this problem by changing their environment in suitable ways. For example, a population of prairie dogs might be faced with frequent extreme fluctuations in temperature; to deal with this, instead of becoming more resistant to temperature changes, they could engineer their environment to be more stable in terms of temperature—for example, they could build and live in an expansive system of underground tunnels (Odling-Smee et al., 2003, pp. 288–289). This idea has become known as "niche construction": a population of organisms altering its own selective environment (Odling-Smee, 1988; Odling-Smee et al., 2003; Laland & Sterelny, 2006; see also Brandon, 1990).[18]

In other words, defenders of the importance of niche construction suggest that, as far as evolutionary theory is concerned, there are two routes towards achieving high fitness (Odling-Smee et al., 2003, p. 375): either by making the organism's needs match what the world is like or by making the world match what the organism needs. To make this clearer, reconsider the dynamics of evolution by natural selection in light of the idea of niche construction (Odling-Smee et al., 2003, pp. 48–50, 239, 370–371; Okasha, 2005; Sterelny, 2005; Laland et al., 2005).[19]

Assume that there is variation in a population of organisms in relation to some set of traits *F*. Assume also that the organisms differ in terms of two *meta*-traits M_C and M_R: they either leave the fitness values of the elements of *F* as they are (the "conservative" strategy M_C), or they engage in (possibly costly) activities that change these fitness values themselves— e.g. by altering the structure of the (physical or social) environment in certain ways (the "radical" strategy M_R). Given this, when concentrating on natural selection, one should expect to evolve whatever feature is fittest—i.e. whichever feature brings organisms the greatest expected reproductive success (Sober, 2008, 1984; Brandon, 1990; Godfrey-Smith, 2009). In particular, in line with (NS) laid out earlier, one should expect a conservative solution to an adaptive problem (i.e. trait M_C) to evolve if and only if

(1) $w(F_i) > w'(F_j)$, for some F_i and all F_j in *F*

and a radical solution (i.e. trait M_R) to evolve if and only if

(2) $w'(F_i) > w(F_j)$, for some F_i and all F_j in *F*

where $w(x)$ is the fitness of trait x in the unchanged environment, and $w'(x)$ is the fitness of trait x after it has been altered by M_R (Odling-Smee et al., 2003, pp. 41–42, 236).[20]

In other words, in this reading, one should expect a population of organisms to pursue a conservative solution (i.e. trait M_C to spread in the population) if it is better (in fitness terms) to leave the fitnesses of the F_k untouched and not engaging in potentially costly niche construction. Conversely, one would expect a population of organisms to purse a radical solution (i.e. trait M_R to spread in the population) if it is better (in fitness terms) to change the fitness values of the F_k and engaging in potentially costly niche construction (Odling-Smee et al., 2003, pp. 301–302). For example, if a type of organism can achieve a higher fitness by remaining susceptible to environmental temperature fluctuations and taking certain steps to dampen these fluctuations (as compared both to remaining susceptible to environmental temperature fluctuations in an environment *with* temperature fluctuations and to changing internally so as to become more temperature fluctuation resistant), we would expect that the relevant activities of environmental temperature smoothing—such as tunnel building—will spread in the population (Odling-Smee et al., 2003, pp. 41–42, 65, 116).[21]

In short, one of the key elements of the niche constructionist approach is the claim that niche constructors can change the evolutionary process itself—they can affect the fitnesses of various traits (of their own or of other organisms) (Odling-Smee et al., 2003). Now, for present purposes, the key point to note concerning this claim is that, purely theoretically, it is entirely uncontroversial: given equations (1) and (2), there is no reason intrinsic to evolutionary theory to think that niche construction—the evolution of radical solutions to adaptive problems—will be any less important than the evolution of conservative solutions (Odling-Smee, 1988; Odling-Smee et al., 2003, chaps. 2 & 3; Laland & Sterelny, 2006; Sterelny, 2003, 2005). Of course, it might be that there are empirical reasons for why (1) is more often satisfied than (2); however, this will then be a contingent fact about the world, and not a principled matter of evolutionary theory. (Also, so far, no clear candidates for these empirical reasons have been found—in fact, the opposite is more likely to be true; see e.g. Odling-Smee et al., 2003, chap. 2.) Hence, at least on the face of it, recent work in evolutionary ecology is right to stress the importance of niche construction: none of the ways of achieving a high organism/environment fit is suspect. As far as evolution by natural selection is concerned, the truth of equation (2) is just as unproblematic as that of (1). This is an important point to keep in mind for what follows in the next section.

2. Adaptive Preferences and Preference Maximization

From the point of view of classical economic theory, it is envisioned that an agent solves a decision problem by picking *actions*: she chooses that action from the set of available ones that lead to outcomes most in line with her preferences. In other words, classical economic theory claims that the agent aims to make the world turn out most like she wants it to turn out. For example, if an agent contemplates buying a new car using her accumulated savings, classical economic theory says that she will choose that car and savings bundle that is highest in her preference ranking (Savage, 1954; Hausman, 2012).

In principle, however, this is not the only way for the agent to achieve a high degree of preference satisfaction: it would seem equally possible for her to simply change her preferences so that they match whatever the world happens to be like. For example, the agent could also make herself prefer keeping her current car and all of her savings over buying a new car and spending some of these savings—that way, she would also end up in a state where the world is the way she wants it to be.[22] Scenarios of this type have become known as being based on "adaptive preferences" (Elster, 1983; Bovens, 1992; Nussbaum, 2000; Zimmerman, 2003; Bruckner, 2007; Hill, 2009; Rickard, 1995). These are situations in which agents change their wants to make them fit to whatever the state of the world happens to be, rather than change the state of the world to make it fit to their wants.[23]

Importantly, however, while both of these ways of achieving high preference satisfaction lead to an agent's preferences and the state of the world being in agreement, for many philosophers (and some economists), these two ways should not be seen to be theoretically on par (Elster, 1983; Bovens, 1992; Nussbaum, 2000; Zimmerman, 2003; Bruckner, 2007; Sen, 1995). In fact, according to many scholars, only the first (world-based) way of achieving preference satisfaction is in line with genuine economic agency; adaptive preferences are generally not even considered to be within the purview of the latter (Sen, 1995, 1985; Nussbaum, 2000).[24] What is less clear, though, is exactly why that is so. What, according to classical economic theory, makes accepting the state of the world and adjusting one's preferences to make them a better fit to that state so different from accepting one's preferences and making the state of the world a better fit to them?

To answer this question, it is useful to reconsider the case of adaptive preferences in the same framework that was employed earlier to describe cases of niche construction. Assume that an agent can do various actions A_k out of a set of actions A. Assume also that the agent has some preference ordering over these different A_k. Finally, assume that the agent can pursue two kinds of meta-strategies: she can leave her preference structure as it is (the "conservative" strategy M_C), or she can change that preference structure in a given set of ways (the "radical" strategy M_R).[25] Again, falling back on using utilities as conceptualizations of these preference structures (again, without loss of generality), it seems that, according to (EA), it follows that we should expect M_C to be chosen if and only if

(3) $u(A_i) > u'(A_j)$, for some A_i and all A_j in A

and M_R to be chosen if and only if

(4) $u'(A_i) > u(A_j)$, for some A_i and all A_j in A

where $u(x)$ is the agent's utility assignment to action x before any possible changes in her preference structure, and $u'(x)$ is the utility assignment to action x after it has been changed by M_R.[26] In other words, it seems that, in parallel with niche construction in evolutionary biology, it should be an upshot of the classical economic theories of choice that agents change their preferences in certain ways (or at least act as if they did so) if that will lead to a higher degree of preference satisfaction, all things considered, than sticking to the preferences they have and trying to find a way to make the world match them—and vice versa.

However, there is reason to think that this is in fact *false*. The classical economic theories of choice do *not* allow the derivation of (3) and (4) as suggested previously. This is because, given how preference values enter into these theories, neither (3) nor (4) are meaningful within these theories at all. A fortiori, therefore, as they stand, these equations—and thus the adaptive preferences that they model—cannot be derived from these theories.[27]

To see this, note that, within the classical economic theories of choice, utilities do not attach to actions outside of a specific evaluative structure. They are extremely weakly cardinal—or even merely ordinal—representations of a given preference profile, and there is no theory-inherent reason for using the same utility-scaling for different preference profiles. (Much the same goes for other conceptualizations of preferences values, such as prospects or regrets; see Chapter 4 for more on this.) For this reason, utilities are not comparable across different preference structures; the different structures can be given very different kinds of utility representations. For example, in von Neumann-Morgenstern expected utility theory—probably the most common classical theory of choice used in economics (Mas-Colell et al., 1996)[28]—utility is only cardinal in first differences: if $u(x)$ is a function assigning utilities to the options in an agent's preference ordering, then all positive affine transformations of $u(x)$ are equally acceptable utility functions representing that same set of preferences (Luce & Raiffa, 1957; Jeffrey, 1983; Mas-Colell et al., 1996). Importantly, while this weak cardinality thus puts some slight limitations on the number of utility functions allowed to represent any given preference structure, it does not alter the fact that the number of possible utility functions is still so large as to make the comparison of utility values across different preference structures meaningless (the only thing that is comparable are ratios of utility differences). In particular, the utility functions used to represent different preference structures can still involve drastically different scalings and thus fail to provide any kind of basis on which to make comparison of utility values across preference structures.[29]

Importantly, this point about the non-comparability of utilities across preference structures holds whether these preference structures belong to different people or to different time slices of the same person. All that matters is that there are different preference profiles—i.e. different utility functions[30]—in play in the situation in question.[31] In turn, what this implies is that this "radical" strategy yields utility assignments that need not be comparable to their unchanged prior state: the utility values $u(x)$ and $u'(x)$ on either side of (3) and (4) do not need to have the same "scaling." Hence, as such, these are not statements that can be made within the classical economic theories of choice—and they therefore cannot be seen as straightforward consequences of these theories.

Now, of course, it is possible to add something to the classical economic theories of choice to make the two utility functions comparable—there is nothing in these theories that precludes this (Harsanyi, 1977; Bradley, 2008; Binmore, 2009).[32] One possibility here is to appeal to a "0 to 1" rule (Hausman, 1995): the least preferred option is given utility 0, the most preferred option utility 1, and all the rest are equally spaced out in between. Various other suchlike proposals exist.

For present purposes, the key point to note concerning these additions to the classical economic theories of choice is that they are differentially plausible in different cases, and there will be some cases where no such additions are plausible (Hausman, 1995). Since different utility functions can represent different things—moral values, physiological states, etc.—different rules for doing inter-/intrapersonal utility comparisons will be plausible in different circumstances.[33] For example, appealing to the "0 to 1" rule will be plausible if the relevant preference profiles reflect an overall similarity in how many degrees of "goodness" there are, for this justifies giving the top and bottom option the same value in both cases. However, this assumption will not always be true. Sometimes two different agents—or one agent before and after she has changed her preferences—can be assumed to evaluate the world so differently that what they prefer most or least differs in conative value, hence making the "0 to 1" rule less compelling.

All of this matters for two reasons here. First, it implies that economists or philosophers are right to not see adaptive preferences as being theoretically unproblematic. In particular, agents that change their preferences to make them match the world need not be maximizing their expected utilities as laid out in (EA)—for classical economic theory, as such, does not assess choices *across* different utility functions.

Second, though, economists and philosophers need not dismiss all cases of adaptive preferences as problematic either. In particular, they are now free to draw a distinction between "good" and "bad" kinds of adaptive preference changes, where the former are in line with classical economic agency *extended with principles of utility comparability* and the latter are not. Indeed, many of the classic discussions of adaptive preferences find a comfortable home in the present theoretical framework. For example, Elster's (1983) claim that adaptive preferences are only defensible if they have been arrived at through conscious, intentional deliberation can be seen as an attempt to spell out when it is possible to make the relevant kinds of utility comparisons: when a preference is arrived at as the result of conscious, intentional deliberation, then that preference is comparable to its initial state (similar remarks can be made about other ways of justifying the defensibility of some adaptive preferences—e.g. those of Bovens, 1992; Rickard, 1995; Zimmerman, 2003; Bruckner, 2009).[34]

These two conclusions concerning the theoretical acceptability of adaptive preferences are inherently interesting. However, beyond this, they are important to note, as they bring out some further differences between the notions of fitness and utility—and thus a further set of restrictions in claim (b).

3. Fitness, Preference, and Differences in Scale

There is a great difference between (1) & (2) and (3) & (4): while the former two can be derived perfectly validly from the theoretical framework of modern evolutionary biology, the same is not true for the latter two and the theoretical framework of modern economics. The latter two can only be derived in certain restricted circumstances. Interestingly, it is also not too difficult to diagnose the *formal* reason for this asymmetry in what can be derived from the two theories. Fitness is normally seen as an inherently *probabilistic* notion—it concerns the expected number of offspring an organism has and may be modulated by a function of the variance of that number (Gillespie, 1977; Brandon, 1990; Sober, 2001). By contrast, preference values, at least in their basic form, are not probabilistic in any way—they might combine with (subjective) probabilities to determine an agent's choices, but they themselves are not probabilistic (Jeffrey, 1983; Hausman, 2012). This fact matters here, as it accounts for why fitness values can typically be changed and still remain comparable, while preference values typically cannot. From the get-go, there is more structure in the former—they are inherently less freely scalable than preference values—and hence they retain meaning even across different fitness assignments. In what follows, I will therefore say that fitness values are *formally richer* than preference values.

Of course, this then raises the question of *why* preference values (at least inherently) are given a less formally rich representation than fitness values. Why do economists accept, at least very often, a less strongly structured account of preference values than evolutionary biologists do of fitness values? In the main, there are two reasons for this.[35]

First, there is a difference in what economists want to do with the notion of preference as compared to what evolutionary biologists want to do with the notion of fitness. In particular, while it is one of the main goals of evolutionary biological analysis to determine which traits can be expected to evolve *in a given time period*, economists often just want to know which strategy has the highest preference value.[36] Importantly, in the latter case, we do *not* need to know how much "better" an option is than some other option—something we do definitely need to know if it is our goal to calculate the speed of convergence towards the relevant stable state. For this reason, economists can be satisfied with a less constrained formal account of preferences as compared to evolutionary biologists' account of fitness.

The second reason why preference values have been given a less rich formal structure than fitness values is that it is not so clear, for every case, what the appropriate richer structure would be for the preference values (if any); equally, it is not clear that there is *one* relatively rich such structure that would be plausible for all cases. This is because—as noted earlier—differences in preference values can have many different sources, and it is not clear that one and the same formally rich structure would work for all of these different cases. So, for example, it may be that, when it is reasonable to see preferences as being based on (expectations of) felt pleasures (Morillo, 1990; see also Glimcher et al., 2005), they can be scaled using a measure with a fixed zero (e.g. the "rest state" of the brain's pleasure centers) and a fixed upper bound (e.g. the maximal level of activation of the relevant brain areas). By contrast, if preferences are based on the number or importance of the goals achieved by an action, a more appropriate measure might be an open-ended scale with a fixed zero and unit (see e.g. Savage, 1954 for views related to this idea; Jeffrey, 1983). In yet other cases—e.g. involving the agent's moral commitments—it might be completely unclear what the relevant measure is (Hausman, 1995). Given this, it may well not be possible to opt for *one* richer account of preference values.

Note that this is different in the case of fitness. The latter always captures the tendency or disposition of an organism to have offspring.[37] To be sure, there are many different "bases" for these dispositions, but, given that they all concern reproductive success, it is not unreasonable to think that they can all be captured by the same sort of formal framework. Put differently, while there might be many different *reasons* for a given fitness difference in a number of organisms, what the fitness difference *represents* is always the same: a difference in the propensity (or some such) of the organism to have offspring. This, though, is not the case when it comes to preference values, where the same preference value difference can *represent* very different things—e.g. differences in the expectation of felt pleasure or in the commitment to a non-personal cause (Hausman, 2012). For this reason, it is much harder to come up with formally richer accounts of preference values than it is for fitness values.[38]

Summarizing all of this, it thus becomes clear that fitness is a formally richer notion than preference because it is substantively narrower than the latter—it captures a smaller set of features of the relevant organism. This is important also because this formal richness is necessary for the work of evolutionary biologists. By contrast, preference is substantively wider than fitness, which makes it harder to give it a formally rich treatment. This, though, is not as problematic in economics, as, in many cases, a formally richer notion of preference values is not needed.

4. Another Restriction in Scope

Of course, this does not mean that the difference in the formal richness/ substantive width of fitness and preference values must always be relevant. Some cases of evolutionary biological and economic analysis only involve one fitness or preference value function and do not consider issues such as the speed of convergence to the relevant optimal value. The point to be noted is just that, if we want to appeal to claim (b) in connecting work in evolutionary biology and economics, we must have reason to think that, in the case in question, comparisons across and within fitness/ preference values are either not necessary, or that all the preference values in question rest on features of an agent's moral outlook, physiology, etc. that permit the notion to have a formal richness that matches that of fitness functions. Another way of putting this point is that any ordinal equivalence between fitness and preference values needs to hold not just synchronically and for the same agent, but also diachronically and across agents.

In this way, we can generalize a point sometimes noted in the literature. In considering the potential uses of evolutionary game theory in economics, for example, it has sometimes been noted that this use is hampered by the fact that preference values, unlike fitness values, are not, in general, interpersonally comparable (Alexander, 2007, 2009; Okasha, 2011; Gruene-Yanoff, 2011). What the previous discussion shows is that the problems for connecting evolutionary biology and economics using a natural selection/economic choice parallelism run deeper than this. It is not just that fitness is interpersonally comparable, and utility is not (in general). The real problem is that in general fitness is substantively narrower and therefore formally less freely scalable than preference (and that is so on all of the major ways of spelling out the nature of preference in economics). Because of this, connecting the two can fail even if issues of interpersonal comparability of utilities are excluded. Even if we are only looking at one agent in isolation, and even if we consider other non-utility-based conceptualizations of preferences, we ought to be careful in applying a natural selection/economic choice parallelism.

We thus arrive at another restriction in the scope of claim (b). For fitness and preference values to function in the same way in evolutionary biology and economics, it needs to be the case that the two notions are scalable to the same degree. With this in mind, it now becomes possible to bring the discussion of this chapter together.

VI. Reprise—The Natural Selection/Economic Choice Parallelism and the Explanation of Attitudes Towards Risk

The question at the heart of this chapter is whether and when it is possible to transfer insights from evolutionary biology to economics by appealing to the fact that the two subjects study the operation of the same kind of constrained optimization process over effectively the same quantity. Putting sections III–V together, we now arrive at the following answer to this question. A structural parallelism between natural selection and economic choice holds if, in the case in question, four conditions are satisfied:

(i)   It must be reasonable to see an agent's preferences as sufficiently strongly correlated with what it is adaptive for the agent to do.
(ii) Economic agency must be seen to be based on preference maximization.
(iii) The difference between global and local maximization must not be relevant for how natural selection and economic agency work.

(iv) Preference values must be able to be seen to have a formal structure
      that is rich enough to allow for comparisons across preference val-
      ues *or* only one preference/fitness value function is appealed to.

Condition (i) is equivalent to claim (a) of section II, and implies that
fitness and preference values are ordinally equivalent. Conditions (ii)–(iv)
are jointly equivalent to claim (b) of section II, and imply that fitness and
preference values are playing the same roles in the core processes studied
in the two subjects: condition (ii) ensures that the economic decision-
making is an optimizing process, condition (iii) ensures that it is an optimizing process of the same kind as natural selection, and condition (iv)
ensures that preferences are based on specific, stable features of the world
that can be meaningfully reassessed if needed.

With this in mind, it is possible to return to section II's illustrative
application of the natural selection/economic choice parallelism surrounding human attitudes towards risk. Recall that this application concerned the question of how to best explain human attitudes towards risk:
as the result of standard preference maximization with non-linear utility
functions, as the result of economic decision-making that is not based
on preference maximization, or as the result of the operation of choice-
external influences on economic decision-making. Recall also that an
appeal to claims (a) and (b)—i.e. the natural selection/economic choice
parallelism—underwrites a combination of the first and third responses.
We ought to expect humans to be fundamentally risk-averse, and any
departures from this risk aversion to be due to choice-external influences
on their decision-making.[39] However, is the appeal to claims (a) and
(b) justifiable in this context?

Given conditions (i)–(iv), it appears that the answer to this question is
mixed. In particular, the discussion here suggests that we cannot appeal
to claims (a) and (b) to underwrite *all* aspects of human attitudes towards
risk. Rather, this evolutionary economic answer to the explanation of
human attitudes towards risk will only be plausible for *some* decision
situations.

In particular, as will also be made clearer in Chapter 4, there are a
whole host of decision situations in which preference maximization does
not appear to be what underlies economic decision-making. For example,
a number of consumption decisions appear to be based on simple heuristics instead (see also Schulz, 2018b). Further, a number of other decision
situations (such as who to marry), while possibly based on preference
maximization, do not feature scalable utilities, as they involve complex
moral or personal commitments. Hence, human attitudes towards risk in
these contexts are unlikely to be well analyzed using the natural selection/
economic choice parallelism either.

However, this does not mean that there are no decision situations that
are justifiably approachable with a natural selection/economic choice

parallelism. In particular, repeated, interactive decisions involving concrete, basic goods seem, at least prima facie, a good case. So, when analyzing how two or more socially similarly situated humans divide food resources in a strategic setting that is repeated, a natural selection/ economic choice parallelism-based analysis of their attitudes towards risk appears—at least on the face of it—reasonable. (For a concrete example of this sort of case, see e.g. Alvard, 2003.)

To see this, note that the choice of foodstuffs is obviously biologically very important (see also Okasha, 2011). Therefore, condition (i) is plausibly satisfied here. Similarly—and as will also be made clearer in Chapter 4 (see also Schulz, 2018b)—repeated, strategic interactions are reasonably seen as based on preference maximization. This satisfies condition (ii). Condition (iii) can be assumed to be satisfied here, as the operation of heuristics and biases constraining the choice set (as hinted at in section III) plausibly makes people only local maximizers. Finally, condition (iv) is plausibly satisfied here, as the division of foodstuffs is, at bottom, about satisfying basic human needs. This makes it plausible that a change in tastes—e.g. away from potatoes and towards plantains— leaves the relevant preference values comparable. Altogether, this makes it plausible that this situation can be well approached using a natural selection/economic choice parallelism. In turn, by the reasoning sketched in section II, we should therefore conclude that these kinds of repeated, interactive food-sharing decisions are inherently risk-averse—with any divergences from this speaking to choice-external influences on economic decision-making.

In this way, the appeal to the natural selection/economic choice parallelism can help make progress in our understanding of human attitudes to risk—though only in some circumstances. It is worth emphasizing that the details of these examples may need to be revised as further insights into human economic decision-making become available (see Chapters 4 and 5). What matters here, though, is just that appeals to a natural selection/economic choice parallelism in resolving open economic questions—such as the one concerning the explanation of human attitudes towards risk—will be differentially plausible in different contexts. While there is every reason to think that some such applications will be entirely defensible, this cannot be taken for granted, and needs to be established on a case-by-case basis.

VII. Conclusions

The postulation of a natural selection/economic decision-making parallelism can be plausible and deserves to be explored—but this postulation involves more complexities than is often realized. Specifically, while not untenable, the ideas that (a) fitness and preference values are ordinally equivalent, and (b) fitness and preference values play the same roles in evolutionary biology and economics, are only plausible in a restricted set of circumstances. However, while restricted, this set of these circumstances is not empty. In particular, the analysis of human attitudes towards risk in repeated, strategic interactions involving the division of basic foodstuffs (at least plausibly) profits from an appeal to a natural selection/economic choice parallelism. In this way, I conclude that the structural project of evolutionary economics focused on a natural selection/economic choice parallelism can be plausible and epistemically important—but only in some circumstances.

Suggested Further Reading

Hammerstein and Stevens (2012) is a wide-ranging collection of essays taking an evolutionary biological perspective on economic decision-making. The papers in Okasha and Binmore (2012) similarly consider a number of the issues raised in this chapter (as well as in Chapter 4), but also discuss other topics such as team reasoning and cooperation. A. J. Robson (2002), Gintis (2009), Okasha (2011, 2007), and Orr (2007) are sophisticated studies of some of the aspects of the natural selection/economic decision making parallelism. Cooper (2001) is an extended investigation of the extent to which logical reasoning can more generally be given an evolutionary biological foundation.

Notes

1. This is not to say that these two theories have historically been developed in tandem—indeed, the opposite seems to have been the case.
2. As will be made clearer in the next section (and in Chapter 4), the extent to which the core processes of choice in economics in fact are optimizing and preference-based is a point of some contention. However, there is no question that the assumption that they are is widely made and has considerable support (at least prima facie). This is all that is needed to get the discussion here off the ground. See also the following sections.
3. Note that the term "adaptive preferences" is sometimes reserved for "bad" (theoretically unacceptable) kinds of adaptive preference changes, and the term "character planning" for "good" (theoretically acceptable) ones (Elster, 1983; Bovens, 1992; Bruckner, 2007). I return to this distinction later; for now, it is just important to note that I shall be using the term "adaptive preferences" in a wider sense that comprises both good and bad versions.
4. It is possible to link evolutionary biology and economics using either just claim (a) or just claim (b). However, this would make for a far weaker connection that is better seen as an instance of the evidential or the heuristic forms of evolutionary economics. Note also that this way of setting up the natural selection/economic choice parallelism cross-cuts the subset selection/ successor selection distinction of Hodgson and Knudsen (2010, pp. 94–104). The idea is that picking the decision option with the highest preference values will be selected for. See also Chapter 5.
5. For more on the nature of risk in economics, see next section and Chapter 5. See also Okasha (2011, 2007).

6. As will also be made clearer in the following sections and in Chapter 4, the core claims of economic theory can be read normatively or descriptively. This same dichotomy also applies here: on the one hand, we can ask whether it is reasonable to expect that people, as a matter of fact, are—or act as if they are—risk-averse, risk-neutral, or risk-loving, and on the other, we can ask whether it is a demand of rationality that people are—or act as if they are—risk-averse, risk-neutral, or risk-loving. This difference is not so important here, though, so I will not discuss it in more detail.

7. This can also enable economic decision making in cases where expected utilities cannot be defined (see e.g. Alexander, 2012). Note also that Allais's distinction between attitudes towards wealth and attitudes towards risk makes a similar point: see e.g. Okasha (2011).

8. A quick word on how this analysis differs from that in Okasha (2011). By considering the work of A. J. Robson (1996), Stearns (2000), A. I. Houston et al. (2007), Curry (2001), and Sinn (2003)—among others—Okasha argues that it may well be plausible that "producing creatures that care about their absolute number of offspring (or things that promote it such as food, sex, etc.) is the best that evolution can do, even though it would be better to produce organisms who cared directly about their relative number. If that is so, then we should expect to see violations of EU maximization" (Okasha, 2011, p. 102). If Okasha is right about this, then the appropriate conclusion would be a combination of options *two* and three for accounting for human attitudes towards risk. However, for present purposes, this is not central. On the one hand, as will also be made clearer in the following sections, the focus on utilities here is purely illustrative, and the present arguments can be reformulated in line with Okasha's defense of non-utility-focused views of economic decision making. On the other hand, in the relevant regards, the upshot of Okasha's analysis is anyway the same as the one here: namely, that economic decision making should be expected to be risk-averse, with departures from this inherent risk aversion being due to other biological, cognitive, social, or other processes interfering with the mechanisms underlying economic decision making. At any rate, the rest of this chapter aims to make clear that on either reading of the situation there is another concern with linking natural selection and economic decision making—namely, the scalability of fitness and preference values.

9. Of course, this does not *entail* that these preferences will evolve: there is more to evolution than natural selection. However, it is also the case that determining that a trait is being selected for provides at least a reason for thinking that it will evolve (Godfrey-Smith, 2001; Orzack & Sober, 1994; Schulz, 2018b).

10. Much the same goes for Okasha's (2011) suggestion that an organism's preference are, for one reason or another, merely an imperfect proxy for what is biologically advantageous for it. See also note 8.

11. Of course, it is possible that social learning has evolved despite being maladaptive. However, given the complexity of this trait, this is relatively unlikely (Dawkins, 1986; Godfrey-Smith, 2001).

12. Another example of this sort of case may be political preferences. A preference for conservative over progressive politics—or the reverse—may not appear to be adaptive, but Thornhill et al. (2009) argue that it may in fact be an adaptive response to the relative amounts of pathogens in the environment. Of course, whether this so deserves further scrutiny, but it again shows that the adaptive value of a given preference need not be obvious.

13. This point can also be made using the distinction between ultimate and instrumental preferences. However, since this distinction is far from clear—see e.g.

Stich (2007), Goldman (1970), Sterelny (2003), and Schulz (2018b, chap. 6)—I phrase the issues here more generally.

14. In what follows, I shall switch between talking about the fitness of traits and that of organisms. While, in general, these two are not interchangeable (Sober, 2000, chap. 3), for present purposes, conflating these is not problematic—I focus on only one trait per organism, so that here the two come out to be the same.

15. Sen (1997) has also argued that we need to distinguish optimization from maximization, and that economic agency should really only be seen to be about the former. However, the present objection goes even further than this, and argues that not even optimization is what is relevant to economic agency.

16. Another reason for doubting claim (b) has been suggested to lie in the fact that fitnesses are interpersonally comparable, while preference values are not (Alexander, 2007, 2009; Okasha, 2011; Gruene-Yanoff, 2011). As will be made clearer in what follows, I think this criticism indeed gets at important issues, but stops short of fully illuminating these issues. Bringing this out is the aim of the rest of this chapter.

17. Interestingly, though, these two phenomena differ in that niche construction involves changes to the world (where the standard cases involve changes to the organism), whereas adaptive preferences involve changes to the agent itself (where the standard cases involve changes to the world). This is not so relevant here, however.

18. Odling-Smee et al. (2003) use the term "niche construction" to widely include all kinds of alterations of the environment—including migration into a new environment, metabolizing food, and photosynthesis. As will become clearer in the next section, though, in this context, I use the notion in the narrower sense (see also Okasha, 2005; Laland et al., 2005; Griffiths, 2005; Sterelny, 2005).

19. What follows is an extremely simplified and abstract reconstruction of these dynamics. For more detailed mathematical models of these sorts of cases, see Odling-Smee et al. (2003, chap. 3 and the relevant appendices).

20. Technically, (1) and (2) would need to quantify over all the available ways of changing the environment (i.e. over all the different available ways of changing w (x) into some different w'(x)). For simplicity, I have left this out here. Note also that the situation here could also be captured by thinking of a matrix where the columns list all of the relevant traits, and the rows the different available fitness assignments to these traits, with the first row containing the current, unchanged fitness values. Then a conservative solution would consist of cases where the maximal value in this matrix is in the first row, and a radical solution in cases where the maximal value is in a row other than the first one. Finally, defining the fitnesses of M_R and M_C themselves is a little tricky, but a natural way of doing so would be by setting $w(M_C) = \max [w(F_i)]$ and $w(M_R) = \max [w'(F_i)]$.

21. There is an alternative way of conceptualizing niche construction (Dawkins, 1982, chaps. 11–13; Odling-Smee et al., 2003, pp. 30, 131–132, 191–192). On this alternative interpretation, one would make a distinction among the elements of F, with some being organism-focused (the F^O_k) and some environment-focused (the F^E_k). Given this, one should expect an organism-focused solution to an adaptive problem to evolve if and only if $w(F^O_i) > w(F^E_j)$, for some i and all j, and an environment-focused solution to evolve if and only if $w(F^E_i) > w(F^O_j)$, for some i and all j (where $w(x)$ is again the fitness of trait (x)). However, for present purposes, this reading of niche construction is less useful than the one in the text; this is so for two reasons. First, the interpretation in the text seems to be more in line with the main idea

behind niche constructionism—namely, that the activities of organisms are both effects and causes of evolutionary processes (Odling-Smee et al., 2003, pp. 48, 112–113, 240; Lewontin, 1982; Sterelny, 2005; Laland et al., 2005; Sterelny, 2003, pp. 148–149). Second, it is especially the interpretation in the text that is useful for clarifying the contrast to the phenomenon of adaptive preferences.

22. More realistic instances of this sort of phenomenon include the fact that in certain circumstances oppressed people (e.g. women) seem to come to prefer a position of little political and social freedoms; equally, it appears that many consumers come to prefer whatever product they happen to own (Sen, 1995; Nussbaum, 2000; Kahneman et al., 1991). Apart from this, there is also much discussion in the medical literature about the fact that patients often seem to change the standards by which they assess their personal situation as that situation changes (a phenomenon known as "response shift"). See e.g. Sprangers and Schwartz (1999).

23. A related class of phenomena concerns cases of so-called "sour grapes" (Elster, 1983; Bovens, 1992; Rickard, 1995; Hill, 2009): situations where agents change the way they think about the world to achieve a higher degree of desire satisfaction (the name refers to La Fontain's fable, in which a fox learns that some grapes he thinks look tasty are hanging too high for him to reach; he then changes his mind and claims that the grapes are—or at least look—sour). This case is different from that of adaptive preferences, in that "sour grapes" involve changes in an agent's *beliefs* (or *perceptions*) rather than changes in her *desires*. In general, cases of "sour grapes" raise different issues from the ones that are being discussed here (Zimmerman, 2003; Bruckner, 2007; Hill, 2009).

24. The existence of adaptive preferences is sometimes also seen as a *moral* issue—in particular, it might be claimed that the existence of adaptive preferences is one of the prime reasons why social policy should not (just) take an individual's preferences into account (Nussbaum, 2000; Sen, 1995, 1985; Hausman, 2012; Nussbaum, 2001; McKerlie, 2007; Bykvist, 2010). However, discussing this further is not necessary for present purposes.

25. Note that there is no clear analogue here to the organism-focused/environment-focused distinction mentioned in note 21. In economics, the actions an agent can do are not normally analyzed in a way that would make it possible to draw this distinction.

26. As before, equations (3) and (4) should also quantify the different feasible ways of changing an agent's preference structure; also, as before, I shall leave this aside for simplicity. Note also that it need not be the case that all ways of changing the utility function are equally feasible: changing one's preferences may be difficult—e.g. in terms of time, concentration, and attention—so that only a few serious options actually exist at most times. Finally, as before, there is a matrix-based alternative way of conceiving this situation. See also note 20.

27. See Welsch (2005) and Hill (2009) for related accounts. Note, though, that Welsch's (2005) model concerns a case in which an agent changes the weights attached to various goals she pursues while keeping the utility she achieves from these weight/goal combinations fixed—which is slightly different from how the issues are set out here. Hill (2009) pursues a line more similar to the one suggested here, though he also does not draw out quite the same implications as is done here.

28. Note also that since circumstances of risk are ubiquitous, using a weaker theory than von Neumann-Morgenstern expected utility theory will not generally be possible either (and, at any rate, utility would then be even

less narrowly circumscribed). I thank Don Ross for useful discussion of this point.

29. So, for example, assuming that B is the gamble of getting A with probability 0.75 and D with probability 0.25, and C is the gamble of getting A with probability 0.25 and D with probability 0.75, then, according to the von Neumann-Morgenstern expected utility theory, it must be that for a utility function u(x), u(B) = 0.75u(A) + 0.25u(D) and u(C) = 0.25u(A) + 0.75u(D) (this is due to the fact that in von Neumann-Morgenstern expected utility theory the utility of a gamble is assumed to be identical to the expectation of the utilities of the gamble's components). In turn, this implies that u(B) − u(C) = 0.5[u(A) − u(D)]; importantly, this will be so for all allowed rescalings of u(x). However, we then still cannot say anything about the absolute values of A through D across the scalings—and hence, across preference structures. So, if an agent starts by preferring A to D (so has u(A) > u(D) for some utility function u(x)), and then changes her mind and starts preferring D over A (so has u'(D) > u'(A), for some utility function u'(x)), we know that her ranking of B and C must also flip (i.e. we know that u(B) − u(C) > 0 > u'(B) − u'(C)). However, importantly, we cannot say that the utility she assigns to B before the change must be higher than the utility she assigns to C afterwards—after all, we could set u(A) = 100, u(D) = 1, giving u(B) = 75.25 and u(C) = 25.75, and u'(D) = 1, u'(A) = 0, giving u'(B) = 0.25 and u'(C) = 0.75. The only statements we can make here concern ratios of utility differences like this one: [u(B) − u(C)] / [u(A) − u(D)] = [u'(B) − u'(C)] / [u'(A) − u'(D)]. The latter, though, are too weak to capture cases of adaptive preferences, as they do not compare the utility values of the options themselves.

30. In principle, differences in preference profiles could also stem from differences in beliefs; however, as made clear previously in note 23, this case raises different issues from the ones at stake here, and thus will not be further considered.

31. The extent to which interpersonal comparisons of utility are similar to intrapersonal comparisons of utility is somewhat controversial; for discussion, see e.g. Broome (1991a), Gibbard (1987), J. Griffin (1986, chap. 6), Hammond (1991), see also Binmore (2009). For present purposes, though, settling these issues is not necessary: no one in this literature thinks that utilities are *always* comparable in the intrapersonal case. This is all that matters here. For some useful discussions of the interpersonal comparability of utilities, see e.g. Bradley (2008), Sen (1979), Hausman (1995), and Goldman (1995).

32. The common extensions of the von Neumann-Morgenstern expected utility theory employed in practice—such as rank-dependent utility or the toleration of an agent choosing strictly dominated options—are *not* sufficient to achieve the kind of comparability needed (Diecidue & Wakker, 2001). This is because these extensions only affect which kinds of transformations of a given utility function are equally acceptable as representations of a given preference ranking; however, they do not lead to the utility values becoming comparable across preference structures. I thank Don Ross for useful discussion of this point.

33. See Hausman (2012) for a defense of the view that preferences in economics can have many different sources. See also Ross (2005), Angner (2018) for a related discussion.

34. The comparability of the relevant preferences does not mean that adaptive preferences *must* be in line with (EA). They would only be so if they did indeed maximize utilities in the case at hand.

35. What follows are not the historical reasons for why a relatively "unstructured" account of preference values was adopted. The point is merely that

there is something to be said in favor of this account (whatever historically led to its adoption).

36. Though not always: see e.g. Busemeyer and Townsend (1993).
37. The nature of fitness is controversial, with some defending a subjectivist, reductive reading of this nature (Rosenberg, 1994, pp. 57–83). However, even these subjectivists about fitness would agree with what follows, but would merely add that these reproductive dispositions will, eventually, be replaced with precise statements of how many offspring each organism has. For more on fitness, see also Sober (2000, 2001), and Pence and Ramsey (2013b).
38. Note, though, that not all fitness comparisons need to be meaningful in biology either; in particular, those across species or taxa might not be. However, the latter failure of comparability would then not be due to the fact that there is no reason to see fitness as having the necessary probabilistic structure, but rather due to the fact that the relevant traits are typically not part of the same evolutionary population (Godfrey-Smith, 2009; Sterelny & Griffiths, 1999, chap. 12). I thank Samir Okasha and Elliott Sober for useful discussion of this issue.
39. Alternatively (as noted in note 8), on the reading of Okasha (2011), this would underwrite a combination of the second and third responses.

3 Market Competition as a Selective Process (The Structural Project II)

I. Introduction

The structural form of evolutionary economics is not restricted to positing and exploring a parallelism between the processes constituting natural selection and those constituting economic choice. Another major set of parallelisms that can be and has been explored switches to the production side of economics—it concerns parallelisms in the operation of natural selection and that of competitive economic markets. Just as natural selection leads to biological entities with higher fitness (assuming heritability) spreading through a population, competitive markets are said to lead to economic entities with higher profits spreading through a market (see e.g. Alchian, 1950; Friedman, 1953; R. Nelson et al., 2018; R. Nelson & Winter, 1982; D. D. P. Johnson et al., 2013; Hausman, 1989). So, rather than focusing on connections in the ways fitness and preference function in evolutionary biology and economics, this form of evolutionary economics focuses on connections in the ways fitness and profits—another core theoretical notion in economics—function in evolutionary biology and economics. In this chapter, I critically analyze the appeal to this kind parallelism further.

To achieve this, I proceed in two steps: first, I lay out some general conditions on what is needed for natural selection and competitive markets to operate in the same way. Given this, I then consider some upshots of the postulation of the exploration of the natural selection/market competition parallelism (where such a postulation is defensible). Note that the structure of the discussion here is different from that of the previous chapter: in the present context, there is more emphasis on the application of the parallelism in question, and less emphasis on the nature of the parallelism itself. This is partly due to the fact that the natural selection/market competition parallelism faces fewer issues than the natural selection/rational choice parallelism—and what issues it faces tend to mirror those faced by the latter, and can therefore be handled relatively more quickly. By contrast, *using* the natural selection/market competition parallelism to obtain insights into open economic questions raises some particularly interesting questions of its own. Hence, this gets a larger share of attention in this chapter.

The chapter is structured as follows. In section II, I lay out general conditions for competitive markets to operate like natural selection, and argue that these conditions are rarely, but still sometimes, satisfied. In section III, I present an open question surrounding the nature of the firm that an appeal to the natural selection/market competition parallelism could possibly help clarify: the question of whether it is reasonable to see firms as agents of their own. In section IV, I present and discuss a model of firm selection in a competitive market meant to address this question. In section V, I asses the actual implications of this model for answering this question. I summarize the discussion and present the overall conclusions concerning the plausibility of appeals to a natural selection/market competition parallelism in section VI.

II. Natural Selection and Competitive Markets: General Considerations

What is a competitive market? A competitive market, as I understand it here, is constituted by the existence of a large number of firms independently producing goods that are relatively close substitutes of each other (Mas-Colell et al., 1996; Knight, 1921). The firms make a profit depending on the amount of goods they sell, the costs of producing these goods, and the price at which these goods sell (Mas-Colell et al., 1996). Firms that do not make a sufficient profit go bankrupt (and thus cease to exist); firms that do make sufficient profits expand in the market. A good example of this sort of process may be the worldwide market for mobile phones. There are a number of different firms in the market that independently produce relatively substitutable phones; they sell these phones at different price points and produce them at different cost points. The extent to which these firms are profitable determines whether they go bankrupt or expand in the market.

An important point to note about this characterization of competitive markets is that it is somewhat more liberal than—though also closely related to—the characterization of (perfectly) competitive markets given in many economics textbooks (see e.g. Mas-Colell et al., 1996; see also R. Nelson et al., 2018, chaps. 3 and 4). So, this understanding does not presume that there are infinitely countless firms in a market that completely take prices as a given. Instead, it is consistent with firms having some pricing power—e.g. because the goods they are producing are not (perceived to be) perfect substitutes of each other (R. Nelson et al., 2018, chaps. 3 and 4). What matters is just that there has to be *some* competition: there have to be a number of different firms, they have to make decisions independently of each other, they have to produce close substitutes of each other, and their survival and position in the market have to depend *only* on their profits. This implies, for example, that markets where the fortunes of firms depend on their having the right kind of connections to the relevant authorities are ruled out. Similarly, the present characterization of markets rules out *systemic* monopolies. If there is only one firm in the market that prevents other firms from entering the market, there is no competition for its goods. However, the characterization does not necessarily rule out all monopolies. If an incumbent monopoly is constantly threatened by the entry of new firms, then a temporary monopoly is consistent with the existence of a competitive market as understood here (R. Nelson et al., 2018, chaps. 3 and 4). Similarly, the present characterization does not rule out all oligopolies. This will depend on whether there are sufficiently numerous different firms in the market that make decisions sufficiently independently of each other to make (differential) profitability a measure of their market power.

Given this relatively broad definition of competitive markets, it becomes possible to start approaching the question of whether competitive markets act like selective processes. Recall that selective processes, as set out in Chapter 1, are constituted by the differential likelihood of replication or reproduction of entities in a population due to differences in their heritable features (see also Godfrey-Smith, 2009). Interestingly, something like this seems to occur in cases of market competition, too. In competitive markets, firms tend to differ in profitability, and profitability is a key determinant of the probability with which firms survive and expand in the market. Hence, over time, we can expect the most profitable firms to take up ever larger shares of a competitive market (see also Friedman, 1953; Alchian, 1950; J. Frank, 2003; Hausman, 1989). In this way, it at least *seems* to be the case that there is a parallelism in the workings of natural selection and market competition. However, things turn out to be more complex than this lets on. Before bringing this out, though, two points in favor of this apparent parallelism need to be noted.

On the one hand, this (apparent) parallelism does not require that firms in fact *maximize* profits; rather, profit *seeking* is all that is required (R. Nelson et al., 2018, chap. 4). This is important, as there are good reasons to doubt that firms do (globally) maximize profits (see e.g. Alchian, 1950; R. Nelson et al., 2018, p. 137; R. Nelson & Winter, 1982; Hausman, 1989). Fortunately, as noted in the previous chapter, for a parallelism with natural selection to hold, *local* profit maximization is all that is needed. The latter, moreover, is not so implausible. While firms may not always succeed in maximizing profits, and while they may use various heuristics in setting prices (R. Nelson et al., 2018, chaps. 3 and 4), it is plausible that they at least *aim at* maximizing profits. This is all that is needed to underwrite this aspect of the natural selection/market competition parallelism (R. Nelson et al., 2018, chaps. 3 and 4).

On the other hand, it is noteworthy that one of the major complexities surrounding the natural selection/economic choice parallelism does not exist in the context of the natural selection/market competition parallelism—namely, differences in the scalability of the relevant notions. In turn, this is due to the fact that profits are inherently more "scalable" than utilities. If a firm initially makes $x profit from producing good A, but then faces the decision as to whether to adopt a new marketing campaign that would allow it to make $y profit from producing the good, it is straightforward to decide whether adopting the marketing campaign is profit maximizing. Profits, unlike utilities, are inherently comparable, both across different firms and across different time slices of the same firm. For this reason, parallels to the problem of the different degrees of "objectivity" of fitness and utility simply do not arise in the context of the natural selection/market competition parallelism.

However, these positive points should not overshadow the fact that there are also restrictions in the scope of the natural selection/market competition parallelism. First and most obviously, not all markets are competitive—even in the broad sense relevant here. Partly by design and partly by accident, many firms produce goods that have few near substitutes (at least in the eyes of the consumer), so that many actual markets simply fail to be captured by this alleged parallelism (Baum & McKelvey, 1999; R. Nelson et al., 2018).

Second, real markets are often more complex than merely involving competing producers and consumers; firms are often both producers and consumers, and are connected to each other in complex ways (R. Nelson et al., 2018). For example, Samsung both competes with Apple in producing cell phones and sees Apple as a consumer of the microchips used in these very phones; both of them are competing buyers of the workers needed to design and make their phones and microchips (Kwok & Lee, 2015). While some of these complexities do not affect the plausibility of a natural selection/market competition parallelism, some of them do. For example, the fact that a firm can be on both sides of an economic interaction with another firm can lead to this firm having an influence in the market that goes beyond its ability to make a profit. Therefore, this issue, too, needs to be kept in mind as a constraint on the scope of this kind of parallelism.

Third, and most importantly, for market competition to be actually parallel to natural selection, the former needs to actually be constituted so as to make for a selective evolutionary environment. In turn, this restriction can be further subdivided into three subsidiary requirements.

(a) Firms need to be able to reproduce, not just grow.
(b) Firms need to differ in features that are inherited by the offspring entities.
(c) The features that are inherited by offspring entities need to be relevant for their profitability.

Condition (a) is needed, as it ensures that the market processes in question are evolutionary processes, not growth processes. While it turns out to be quite difficult to make clear exactly how to draw the distinction between growth and evolution (see e.g. Godfrey-Smith, 2009; Clarke, 2016; Brandon, 1990), and while it may be possible to see some aspects of growth processes be *similar* to evolutionary processes, a genuine *parallelism* between market competition and natural selection requires that it is reasonable to see firms as reproducing, not merely growing (J. Frank, 2003; Brandon, 1990; Hirshleifer, 1977). It is widely agreed that replication is an essential, individuating component of genuinely evolutionary processes (Brandon, 1990; Godfrey-Smith, 2009). Hence, as noted in Chapter 1, a genuine parallelism between market competition and natural selection needs to respect this point—otherwise, this becomes merely a case of analogy (i.e. the heuristic form of evolutionary economics).

Condition (b) is needed, as without heritability, evolution cannot take place: information needs to be transmitted across generations. As has been shown (Godfrey-Smith, 2009; Sober, 1984), this does not mean that the heritability needs to be perfect; however, unless there is some correlation between parent and offspring, evolution has nothing to take a hold in, and cannot take place.

Condition (c) is needed, as selective evolutionary processes require differences in fitness (as also noted in Chapter 1). So, for market competition to be parallel to natural selection—and not to another kind of evolutionary process—whatever is inherited by offspring firms needs to be relevant to their success of replication. By the assumption that the market in question is a competitive market, this implies that whatever is inherited by offspring firms needs to be relevant to a firm's profitability.

Now, it needs to be noted immediately that these are in fact quite demanding conditions. So, for condition (a) to be satisfied, firms need to give rise to genuinely independent offspring firms—i.e. these offspring firms need to operate as independent entities in the market place. This requires firms to have a highly decentralized management approach. In particular, each token of the firm needs to make its own decisions about what to do (including whether to open up further offspring firms), for otherwise, these offspring firms are not genuinely independent entities, and this would be a case of growth, not evolution (J. Frank, 2003; Hirshleifer, 1977; Ghiselin, 1987; Klepper, 2009). Similarly, the kind of heritability called for in condition (b) is only sometimes instantiated. There may well be cases where offspring firms share many aspects of their parental firms, but in many cases, a firm has its own internal dynamic that cannot easily be replicated, and which is therefore not inherited (R. Nelson et al., 2018, chap. 4; Klepper, 2009). Finally, condition (c) restricts condition (b) even further, for it need not only be the case that offspring firms share *some* features with their parental firms, the features they share need to be relevant to their profitability. What this means is that conditions (a)–(c) need to be acknowledged to significantly restrict the applicability of the natural selection/market competition parallelism. It is not just that this parallelism only applies to sufficiently competitive markets that are sufficiently simple, but these markets also have to contain genuinely replicating firms whose offspring resemble them in the right ways.

However, it is also important to recognize that there is no a priori reason to think that these conditions can never be satisfied. For example, while it is true that firms typically seem to make decisions in a more centralized manner than what is assumed in condition (a), it is also true that decentralized decision-making has been found to be both quite efficient and to describe various real firms reasonably well. In particular, some franchising setups could be seen to fit into this model, as well as some large production conglomerates (see also Witt, 2003; Carley, 1996; Gindis, 2009; Klepper, 2009; Radner, 2006). Something similar goes for the other conditions needed to get the natural selection/market competition parallelism off the ground. For example, it is entirely possible that something like a profit-defining "firm culture" is inherited from parental to offspring outlets—e.g. through the fact that the new employees hired by a given set of old employees tend to be similar in key ways to these old employees, or because offspring firms inherit the production, marketing, and sales routines at the heart of the parental firm (R. Nelson et al., 2018; R. Nelson & Winter, 1982; Witt, 2003; Hodgson & Knudsen, 2010). (This issue will become important again in the following section.)

In turn, what this implies is that only a detailed analysis of the particular market in question can reveal whether it satisfies the condition needed to underwrite a natural selection/market competition parallelism. There can be no presumption that this needs to be the case, nor should there be a presumption that it cannot be the case.[1]

Importantly, though, this is not where the complexities surrounding the parallelism between natural selection and market competition end. For even if it is assumed that we are dealing with a market of the right sort—i.e. one that is structured in such a way that the parallelism to natural selection is appropriate—it is not obvious what the appeal to such a parallelism can actually *do* for economists. What open economic questions can this parallelism help address? When it comes to the natural selection/economic choice parallelism, answering this question is not as difficult: as noted in the previous chapter, this has some known applications (Okasha, 2011, 2007; Schulz, 2008; Cooper, 1987, 1989). In the context of the natural selection/market competition parallelism, though, this question is harder to answer—the possible applications of this parallelism are not as obvious.[2] The rest of this chapter thus aims to take steps towards an answer to this question. To do this, it begins by considering a classic and still ongoing debate in producer theory: what *firms* actually are.

III. The Nature of Firms

There are many issues of contention surrounding the nature of firms (see e.g. Coase, 1937; Alchian, 1950; R. Nelson et al., 2018; Schmalensee & Willig, 1989; Radner, 2006; Satz & Ferejohn, 1994; Nickerson & Zenger, 2004; O. Hart, 2008; Enke, 1951; Fama, 1980), but the question at the forefront here concerns what *kinds* of things firms are. Are they genuine economic agents, albeit made up of other such agents, or are they merely spaces within which such agents interact (Knight, 1921; Coase, 1937; R. Nelson & Winter, 1982; Hodgson, 1999, chap. 11; Schmalensee & Willig, 1989; see also Witztum, 2012).[3]

There is no question that this issue has considerable historical importance. Indeed, it is closely related to some of the core concerns that drove the heterodox form of evolutionary economics—namely, the idea that firms are more than transactional spaces (R. Nelson et al., 2018; R. Nelson & Winter, 1982; Groenewegen & Vromen, 1999; Vromen, 2009). However, here, the fact that this issue concerns questions of foundational importance to contemporary debates in economics is more important. Firms are one of the major types of economic actors, and the question of what they are goes to the very heart of much of contemporary economic theorizing. Indeed, this question is embedded in a wide variety of other open economic questions, such as the plausibility and necessity of "methodological individualism" and "microfoundations" (Watkins, 1952; Elster, 1982; Hoover, 2010, 2015, 2001, 2009), the importance of having free markets (Hayek, 1967; Radner, 2006; Schumpeter, 1942; R. Nelson et al., 2018), and the nature of economic agency (R. Nelson et al., 2018; R. Nelson & Winter, 1982). In short, the question of what firms are—and, relatedly, of why there are any firms in an economy to begin with—is a classic but still important dispute at the heart of economics.

For present purposes, three broad approaches of the nature of the firm can be distinguished (Hodgson, 1999, pp. 247–249; Fama, 1980; R. Nelson et al., 2018; R. Nelson & Winter, 1982; Hodgson & Knudsen, 2010; Foss & Klein, 2008; for an interesting legal view of this distinction, see Gindis, 2009). These approaches will be called the "contract-based" approach, the "agent-based" approach, and the "routine-based" approach in what follows.

According to the contract-based approach, firms are a special kind of "transactional space" (Coase, 1937; Williamson, 1971; Klein et al., 1978; O. Hart, 2008; Aghion & Holden, 2011). In particular, on this kind of account, it is posited that there are so called "transaction costs" that come from taking part in a market. These costs include the costs that come from acquiring the relevant information (where which goods and services are bought and sold, at what price, etc.), the costs that come from drawing up the explicit contracts about what is to be bought and sold, and the costs of monitoring and enforcing these contracts. The (posited) existence of these costs is then taken for a reason for removing some economic transactions from the market. Firms can be assumed to grow up to the point at which, for the last transaction in question, the costs that come from taking part in the market are matched by the costs that come from producing the relevant good or service in-house (which include higher input costs, the costs of the acquisition of new skills, etc.). What matters most for present purposes about this approach is that, according to it, firms are a certain kind of economic environment within which economic agents act—they make it possible for economic agents to transact more cheaply than what would be possible on the market. The firms themselves, though, are *not* economic agents—they are merely transactional spaces. As Williamson (1979, p. 239) puts it, firms are just an "institutional matrix within which transactions are negotiated and executed" (see also Kreps, 1990, p. 724).

By contrast, on the agent-based approach to the nature of firms, firms are seen as goal-directed collectives (Knight, 1921, chap. 11; Clark, 1997, chap. 9; Hodgson, 1999; Penrose, 1959). In particular, here, firms are seen as collections of economic agents that have accumulated, on the one hand, a certain set of "quasi-beliefs" that give them the ability to produce and market some particular good or service; and on the other, a set of "quasi-desires" that give them the motivation to see through this production and marketing (Knight, 1921; Hodgson, 1999, pp. 251–258; Simon, 1957; Pettit, 2003). Note that these firm "beliefs" and firm "desires" can differ from those of any of their employees (they are "emergent properties" of sorts). Firms are thought to frequently "know more" than (or "believe" differently from) what their employees know (or believe), and might "want" to do things that none of their employees want to do. As Winter (1988, p. 170) puts it, "it is firms, not the people that work for the firms, that know how to make gasoline, automobiles, and computers." Note also that, while questions can be raised about the differences between firm beliefs and desires and the beliefs and desires that feature in intentional explanations of human actions (see e.g. Schulz, 2018b for more on the latter), a detailed treatment of this point is not necessary for present purposes. All that matters here is that firms are assumed to have beliefs and desires that are sufficiently like the beliefs and desires of individual human agents so that they can act in a goal-directed manner in their own right (leaving open exactly what this implies) (List & Pettit, 2006). Hodgson and Knudsen (2010, pp. 170–171, footnote 9) express this quite clearly: "Organizations as here defined have the capacity for goal-directed behavior, irrespective of whether goals are actually declared. In this sense, an organization has the capacity to be a 'collective actor.'"

Finally, according to the routine-based approach, firms are seen as the embodiment of a set of behavioral routines that determine how a given good is produced (R. Nelson et al., 2018; R. Nelson & Winter, 1982).

These routines may include quite concrete behaviors (e.g. the order in which a product is assembled) and more abstract ones (e.g. how the sales price of the firm's product is set). A firm is seen as successful to the extent that it gets to retain or replicate the behavioral routines that make it up. In this way, the routine-based approach encompasses elements of both the contract-based and the agent-based perspective. On the one hand, firms here are seen as having a genuine (routine-based) identity that goes beyond being an "institutional matrix" within which economic agents can transact, but on the other, firms here are not seen as decision-making, goal-directed systems, but merely as bundles of behavioral routines that operate automatically.[4] For present purposes, though, it is best to see the routine-based approach as a kind of non-agential view. The reason for this comes out most clearly when it is noted what happens when there is conflict as to which behavioral routines a firm ought to adopt. In a case like this, the routine-based view will only track the routines themselves. The firm then again becomes merely a kind of environment for the inter-action of the economically meaningful entities—it is just that, in this case, the latter are behavioral routines (see e.g. R. Nelson & Winter, 1982, pp. 107–112; see also the model that follows). In short, as R. Nelson and Winter (1982, p. 134) put it,

> it is quite inappropriate to conceive of firm behavior in terms of deliberate choice from a broad menu of alternatives that some external observers consider to be 'available' opportunities for the organization. The menu is not broad, but narrow and idiosyncratic; it is built into the firm's routines, and most of the 'choosing' is also accomplished automatically by those routines.

Summing up, one can distinguish two broad opposing viewpoints about the nature of the firm. There are fundamentally non-agential views, which comprise the contract-based and the routine-based approaches; and there are fundamentally non-agential views, which comprise the agent-based approach. For what follows, it is furthermore interesting to note that R. Nelson and Winter (1982, pp. 134–136) analogize behavioral routines to genes, which suggests that their theory of the firm is akin to a "gene's eye" view of biological evolution. Indeed, one could analogize the three different views of the nature of the firm to three different views about the nature of biological entities like colonies of social ants: one could see these colonies as merely assemblies of convenience of its members (the contract-based view), as "superorganisms" (the agent-based view), or as more or less homogenous gene-pools (the routine-based view).[5] I return to these three different views of biological systems in the following sections.

While there is much more that could be said about each of these views of the nature of firms, for present purposes, it is sufficient to note the mere fact of this diversity. All of these approaches to the nature of the firm have their critics and defenders (Knight, 1921; Williamson, 1971; R. Nelson et al., 2018; R. Nelson & Winter, 1982; Witztum, 2012; List & Pettit, 2006; Pettit, 2003). This is not surprising, given how intertwined these issues are with questions surrounding the need for "microfoundations," the importance of market competition, and the nature of economic agency. However, what is key for present purposes is that a root cause of this diversity of perspectives concerning the nature of firms is that it is not yet clear *when* the two major types of views—the agential and the non-agential—are plausible. While it may be agreed that it is *possible* for firms to be agents of their own, it is not yet clear when that is actually the case. For example, it may be granted that considerations like the ones put forward in List and Pettit (2006), Pettit (2003), and Hodgson (1999) show that the agential view of the firm is coherent. However, until it is made clear when exactly that view is plausibly held, these kinds of possibility arguments cannot resolve the debate surrounding the nature of the firm. What needs to be shown, therefore, is when conditions are such that it is appropriate to see firms as agents of their own, and when not. Put differently, given that the agential and non-agential views are not mutually exclusive, the key issue to address here is not whether or not *all* firms *always* are agents of their own. The real issue to address is under what *actual* circumstances *which* firms are agents of their own (which of course leaves open the possibility that there are no such circumstances).

It is also important that it is quite tricky to find ways of answering this question. This is partly because this question is not located within the confines of standard economics; in fact, it is about the nature of these very confines. Partly, this is because this question is related to a number of other foundational economic questions—concerning the need for microfoundations, the importance of competitive markets, and the nature of economic agency—whose answers (and even their formulation: Hodgson, 2007a) is far from clear. Both of these points make it difficult to use the standard tools of economic analysis to investigate the question of what firms are.

Given this, an evolutionary biological perspective becomes quite attractive. Such a perspective could, at least in principle, provide a fulcrum with which to connect the many issues at stake in this context: individual decision-making, group decision-making, and competition. It may be able to give us tools with which to investigate the nature of firms that standard economic analysis on its own cannot. After all, the investigation of individual and group decision-making in competitive circumstances is a key aspect of contemporary evolutionary biology (Alcock, 2013; Sober & Wilson, 1998; Okasha, 2006). For this reason, it is at least conceivable that an appeal to a parallelism in the working of natural selection and market competition can open new inroads into the question of the agential nature of firms. However, whether this is in fact the case remains to be seen. As noted in the previous two chapters (and as will also be made clearer in the remaining ones), the proof of the pudding of the epistemic plausibility of evolutionary economics is in the eating, and needs to be assessed on a case-by-case basis.

IV. Firms, Employees, and Multi-level Market Selection: A Model

To assess whether an appeal to a parallelism in the working of natural selection and market competition can tell us something about the agential or non-agential nature of firms, it is useful to develop and analyze an evolutionary model of firm competition and evolution.[6] Presenting this model is the aim of this section; its interpretation and analysis is left to the next section. I begin by laying out the basic structure of the model, and then present the results of a set of simulations of this model.

1. The Structure of the Model

The model's structure is seen most easily if it is broken down into three stages. First, there is a setting-up stage. It is assumed that there is a competitive and open market for a certain good, that there is sufficient demand for the good to (at least potentially) support a large number of firms, that there are a number of firms that are looking to access that market, and that there are a number of potential employees looking for work in this industry.[7] Formally, there is an x by y grid representing the market, with each spot on the grid being able to hold one outlet; initially, $m \in N$ outlets are created, each of which is assumed to belong to a different firm type, and randomly distributed over the grid (for more on the difference between outlets and firms, see the following section).

Each of these firms/outlets then hires $n \in N$ employees. In turn, potential employees are of two types: they can be hard working employees that go beyond what is minimally required for the job—the "expectation-exceeders" in what follows—or they can just provide this minimum level of effort—the "clock-punchers" in what follows. Firms initially hire expectation-exceeders and clock-punchers in proportion to their overall number in the applicant population (p_E and $1 - p_E$, respectively). At this point, there is no "firm culture" in place that would allow firms to move away from a relatively superficial, random sampling-based search. (I return later to the reasons for making this assumption about the hiring process.)

The second stage of the model concerns what happens within the firms once they are set up. Here, the key thing is that employees can change their type as time progresses—they *learn* during their tenure in a given firm's outlet. More specifically, there is a chance that expectation-exceeders turn into clock-punchers. This chance depends on two factors:

the number of clock-punchers in their vicinity, and the length of time during which employees interact before the next phase of the model begins. Formally, for every expectation-exceeder, there is probability $p_{td} = IP + E_1 d/n$ per learning round that she will turn into a clock-puncher, where $E_1 \varepsilon R$ and in the interval $[0, 1]$, $IP = 0.01$ as a fixed parameter (to ensure that there always a minimal probability for expectation-exceeders to turn into clock-punchers), and d is the number of clock-punchers in the relevant outlet. It is furthermore assumed that there are r learning rounds.

The reason for this way of modelling the situation is that, on the one hand, it can be assumed that being an expectation-exceeder takes time and effort on the part of the employee—i.e. it comes with costs. Furthermore, there are few "external" upsides to being an expectation-exceeder. It can be assumed that firms do not treat expectation-exceeders and clock-punchers differently, as distinguishing these two types may be very difficult for them (for example, both types of employees might work the same number of hours per week—just with different levels of "focus"). Also, it is assumed that there are ample equivalent employment opportunities, so that an individual employee does not care about losing her job.

On the other hand, it is also reasonable to assume that people do not necessarily start out as somewhat "cynical" clock-punchers. The environment they act in is likely to influence them in their attitudes towards the work (see also R. Nelson et al., 2018, pp. 135–136). This is so for several different reasons. First, as was suggested in the last chapter and as will be made clearer in the next chapter, it may take people time to determine what the preference-maximizing choice on offer is (Binmore, 2007; Radner, 2006; Witt, 2003). Hence, it is plausible that the probability of becoming a clock-puncher depends on the number of clock-punchers in the firm (i.e. d/n) and the number of interactions among employees (i.e. r) (Boyd & Richerson, 2005; Sterelny, 2012a). Second, the higher the number of expectation-exceeders in the firm, the more firm culture changes towards a high-effort culture (Henrich, 2015; Nisbett et al., 2001; Fehr & Gaechter, 2000). If there is a high number of expectation-exceeders in a firm, this raises the perceived behavioral standard of all employees. For example, the mere fact that others *can* punish free riders has been shown to increase people's cooperative tendencies (Fehr & Gaechter, 2000, p. 167, figure 3.2). Similarly, therefore, it is plausible that the mere fact that many others work hard can increase people's disposition to work hard themselves (see also R. Nelson et al., 2018, p. 31; Helfat & Campo-Rembado, 2016).

The third and final stage of the model concerns the interactions among firms on the market. These interactions are determined by the *profitability* of a firm. In turn, a firm will be more profitable the more expectation-exceeders there are among its employees. The employees will work more efficiently, creatively, and cooperatively, and thus create a better product (R. Nelson et al., 2018, p. 101). Specifically, firm profitability influences inter-firm relations in two ways. The first concerns outlet survival. Unprofitable outlets are less likely to be able to pay their running costs, cover tax bills, or unexpected expenditures of one kind or another, and thus are more likely to go bankrupt. (As will be made clearer in the following section, less profitable outlets are also more likely to be bought out by more profitable outlets from other firms.) Formally, for every outlet, there is a probability $p_b = E_2 d/n$ that the outlet goes bankrupt, with E_2 ε R and in the interval [0, 1]. The second effect of profitability on the interactions between firms concerns the firm's chances of opening up new outlets (assuming they have not gone bankrupt). More profitable firms have the cash flow to lay out the down payment for a new lease of new premises, start a new hiring process, etc. (Capron & Mitchell, 2009).

To understand both of these effects better, one needs to understand a key distinction made in the model: that among firms and outlets. Firms here are a *type* of organization, the instances of which are its *outlets*. In other words, a "firm" here is merely a higher-order grouping of a set of outlets. The actual businesses "on the ground" are all different outlets (some of which might belong to the same firm). In the model, it is furthermore assumed that outlets have a lot of independence. In particular, every outlet gets to make its own decisions about opening up a new outlet of the same firm type, and every outlet is on its own as far as its continued existence in the market is concerned. The idea, then, is that the managers of the firm as a whole take a very hands-off approach to managing the outlets that make it up; the only instructions that they impart to the managers of a new outlet is that they (the new managers) ensure the new outlet stays in business for as long as possible, and that they open up a new outlet when that is feasible.

As noted earlier, this highly decentralized way of managing outlets is not the most common way in which firms operate. However, as also noted previously, this does not mean that this assumption of a highly decentralized approach is *never* plausible. Hence, this assumption, while restrictive, does not mean that the model has no realistic applications whatsoever. Note also that I do not consider the question of why this particular organizational structure was chosen by the firms' owners here—my concern is with establishing what happens given this particular structure. As noted earlier, the goal is to investigate the *uses* of an appeal to a natural selection/market competition parallelism in economics. For this reason, I assume that firm firms/outlets have the kind of structure required for the natural selection/market competition parallelism to get off the ground at all.

Two more points are important to note about how the present model treats the process of opening up a new outlet. First, opening up a new outlet can be done either by starting one up from scratch, or by buying out less successful outlets of other firms and turning them into outlets of one's own firm (an outlet cannot buy out an outlet of its own firm, as it is not in competition with outlets of its own firm in the same way as it is in competition with outlets of other firms).

Second, the nature of the hiring process is different here from that of the initial outlets. In particular, outlets opening up a new outlet are no longer assumed to sample the applicant pool randomly, but to *replicate the composition of expectation-exceeders and clock-punchers in their own outlet.* The main reason for this assumption is that it is plausible that outlets with more expectation-exceeders will not only be more profitable, but also be better at hiring new employees. The expectation-exceeders will work better and harder together not just in producing the relevant good or service, but also in screening applicants, interviewing them, and making the selection. Hence, they are more likely to hire "high-quality" applicants. In turn, this makes it plausible that newly opened up outlets will tend to look like their "parental" outlets in terms of the composition of their employee types.

For what follows, it useful to comment briefly on the somewhat "Lamarckian" nature of the inheritance process here. The inheritance process in the model seems to involve acquired characteristics—the number of expectation-exceeders in the firm. This, though, does not raise problems for the present appeal to the natural selection/market competition parallelism. As noted in Chapter 1, granting that the present inheritance process is Lamarckian does not invalidate the parallelism to natural selection. *How* traits are inherited is not so central to natural selection—what is central is *that* they are inherited (and exactly how they are inherited is increasingly seen to be diverse and complex) (Godfrey-Smith, 2009; Boyd & Richerson, 2005; Jablonka & Lamb, 2005). For this reason, the fact that the inheritance in this model looks somewhat Lamarckian does not detract from the model being based on a natural selection/market competition parallelism.

Formally, the process of opening up new outlets of the same firm type is modeled as follows. After determining which outlets go bankrupt, the surviving outlets assess whether they are in a position to open up another outlet of the same firm type. This happens with probability $p_{nf} = E_3(1-d/n)$, with $E_3 \in R$ and in the interval $[0, 1]$. If the outlet thus determines that it can open a new outlet of the same firm type, it searches randomly either through the entire grid or through their neighborhood (a parameter that can be changed, but which does not seem to alter the conclusions of the model much), and stops when it has either found an empty spot or one occupied by an outlet of a different firm type that has more clock-punchers among its employees. When a suitable spot for the new "offspring outlet" has thus been found, the new outlet will hire employees in such a way that they match the composition of clock-punchers and expectation-exceeders in the parental outlet.

Overall, the model then consists of repeated cycles of stages 2 and 3: the employees within the various outlets interact with each other and (potentially) adjust their strategies, after which the outlets' profitability is assessed and given the (probabilistic) power to influence (a) whether a given outlet gets to survive and (b) whether it can open up a new outlet of the same firm type. The cycles will end if a given number of outlets exist (i.e. the market is saturated), or the number of existing outlets falls below its initial state m. In the former case, no new outlets can be opened (though old ones can still go bankrupt as before), and in the latter case, a number v of new firms/outlets are created as in stage 1, so as to bring the number of existing outlets back to m. (This is in line with Schumpeterian models of competition, where there are always new firms waiting to enter into a market—see Schumpeter, 1942, p. 84.) Then, the rotation of stages 2 and 3 begins anew.

2. The Results of the Model

It is possible to get a fairly good description of the major patterns in the results of the model by simulating runs of the model under a wide range of different parameter values.[8] To analyze the model, I have thus done the following.

First, I have held fixed the following parameters: the size of outlets (i.e. n) was 10 employees, the initial number of firms (i.e. m) was 10, 50% of employees in the global pool of hirable employees were expectation-exceeders (i.e. $p_E = 0.5$), and outlets could only open up new outlets in their immediate neighborhood. The maximum number of firms a market was assumed to be able to contain was set at 400. Each run of the model lasted 1000 inter-outlet interactions. Given this, I then considered a number of different scenarios. In particular, for three settings of r (1, 4, and 7), I have successively set E_2 at 0.1, 0.5, and 0.9, and then, in each of these cases, let E_1 and E_3 vary from 0.1 to 0.9 (in 0.1 increments). Finally, I have switched E_2 and E_3 and repeated the exercise (i.e. set E_3 at 0.1, 0.5, and 0.9, and then, in each of these cases, let E_1 and E_2 vary from 0.1 to 0.9 in 0.1 increments). This yielded three main patterns, depending on how the different parameters were set.[9]

First, if the rate of firm creation is sufficiently faster than that of firm-internal learning, a firm will eventually establish itself and take over the entire market.[10] Call this the *positive scenario*. A typical plot of eight variations on this theme looks like this (the y-axis represents the number of outlets in the economy, and the x-axis the number of inter-outlet interactions):

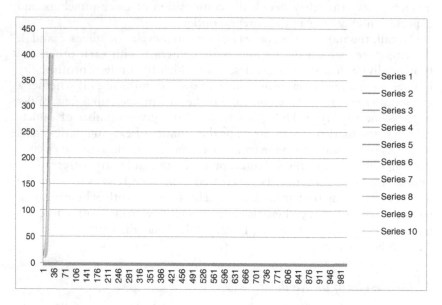

Figure 3.1 A set of eight typical runs in the positive scenario

The second scenario to consider here comes about when expectation-exceeders are turned into clock-punchers relatively easily—i.e. if the rate of firm-internal "learning" is sufficiently faster than the rate of firm creation. Then, genuine market selection of different firms does not take hold. Firms/outlets constantly get created and go out of existence, with none of them having the ability to persist for an extended period of time. Call this the *negative scenario*. It is important to note that this scenario is in fact quite common. In many cases of the model, there is constant firm/outlet ending and firm/outlet creation, without any firm/outlet getting to persist long in the market at all (the total number of firms that is being created here quickly moves into the thousands). A typical plot of eight variations on this theme looks like this (x- and y-axes are as in the previous figure).

Finally, if the rate at which new outlets can be opened is sufficiently balanced by the rate at which individuals learn, a cyclical behavior ensues. One or a few firms temporarily take over the market, collapse, and a new set of firms takes over the market. Call this scenario the *neutral scenario*. A typical plot of eight variations on this theme looks like this (x- and y-axes are as in the previous figures).

It further turns out that, in this model, there are three parameters that are particularly important for the occurrence of a positive scenario (as opposed to a negative or neutral one):

(a) the number of interactions among employees (i.e. parameter r);
(b) the elasticity of outlets going bankrupt on the number of expectation-exceeders (i.e. parameter E_2);

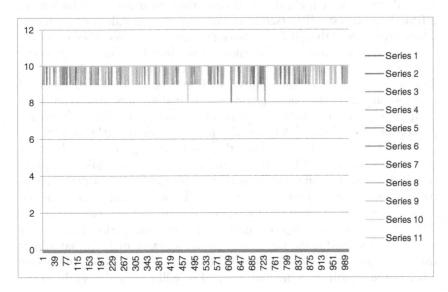

Figure 3.2 A set of eight typical runs in the negative scenario

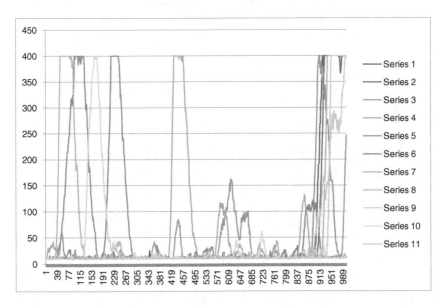

Figure 3.3 A set of eight typical runs in the neutral scenario

(c) the elasticity of outlets opening up offspring outlets on the number of expectation-exceeders (i.e. parameter E_3).

What is furthermore interesting to note is that, among these parameters, E_3 matters much more than E_2. For a positive scenario to occur,

E_3 needs to be quite high; if it is too low, then no matter how high E_2 is, a firm does not establish itself in the market. Higher values of E_2 matter only to the extent that, if E_3 is above the threshold value, they make the occurrence of positive scenarios more likely. In other words, differential outlet replication is much more important that differential outlet survival. This is in fact quite in line with what should be expected from general evolutionary theoretic considerations—survival matters only in so far as it affects reproductive success, whereas the latter affects the evolutionary process directly (Sober, 2000; see also R. Nelson et al., 2018, p. 95).

Another interesting point about this model (though this may be more due to its specific details, rather than being a general finding—a point to which I return momentarily) is that the occurrence of a positive scenario does not depend much on the elasticity of learning of expectation-exceeders on the number of clock-punchers around (i.e. parameter E_1). All that really matters is the length of the learning period (i.e. r)—the higher the value of r, the higher the threshold value of E_3 is that allows the occurrence of positive scenarios. At the extremes, firms either have a hard time establishing themselves at all or will nearly always do so.[11]

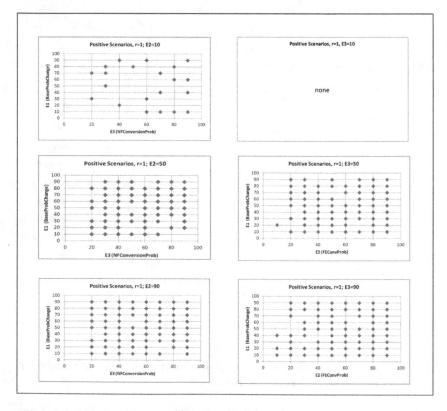

Figure 3.4 Distribution of positive scenarios for r = 1

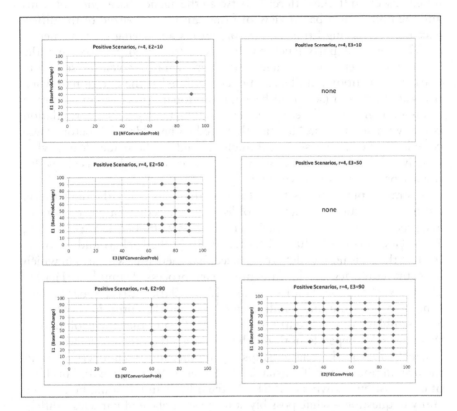

Figure 3.5 Distribution of positive scenarios for r = 4

In a bit more detail, the relationships between r, E_1, E_2, and E_3 in the creation of positive scenarios can be illustrated with the following graphs.

These graphs illustrate four key conclusions of the simulation of the model:

(1) For a given value of r, the proportion of the E_1/E_3 space (if non-zero) that allows for the evolution of stable firms increases as E_2 increases.
(2) If E_3 is less than a given threshold level, positive scenarios do not occur, no matter the value of E_2.
(3) The threshold value for E_3 to lead to the occurrence of positive scenarios increases as r increases.
(4) There were no positive scenarios for $r = 7$.

V. Implications of the Model

What implications does this model have for the question of the agential or non-agential nature of firms? To determine this, recall that, as noted in section II, the difference between the agency-based view of firms and the non-agency-based view of firms can be analogized to the difference between seeing firms as *superorganisms* and seeing them as merely *vehicles* for their parts (such as their genes/routines).[12] Given this, the question of when firms are agents of their own can, given the selection/ market competition parallelism, be reformulated as the question of when firms are targets of (selection-based) evolutionary processes in their own right—though, as will be made clearer momentarily, this reformulation comes with several complexities that need to be addressed. Before these complexities can be assessed, though, it needs to be made clearer what exactly it takes for something to be a target of (selection-based) evolutionary processes in its own right.[13]

Abstractly put, it needs to be the case that it is really the higher-level entities in question that are the objects the (selection-based) evolutionary processes range over, rather than their parts (Sober, 1984; Brandon, 1990; Godfrey-Smith, 2009; Clarke, 2016). In the background of this is the fact that it is now widely recognized that there are two ways in which an entity could feature in an evolutionary process (Damuth & Heisler, 1988; Okasha, 2006; Godfrey-Smith, 2009; Clarke, 2016; Sober & Wilson, 1998). On the one hand, it could be that the entity in question is an assemblage of other entities—"particles"—and that the evolution of these particles depends both on the nature of the other particles in the assembly *and* on the relationships *among* different such assemblages in the overall population (Okasha, 2006; Sober & Wilson, 1998). This kind of case is often known as MLS1. On the other hand, it could be that the entity in question—while possibly still an assemblage of particles—differs in some *assembly-level* traits from other such assemblages, which moreover confer fitness benefits *to that assembly*, and which are heritable *at that level* (Okasha, 2006; Godfrey-Smith, 2009; Clarke, 2016). This kind of case is often known as MLS2.[14]

For present purposes, the key point about this distinction is that for firms (or anything else) to be genuine *targets* of (selection-based) evolutionary processes, they have to evolve in this second, MLS2-based way. In the first, MLS1-based case, what evolves is really only the particles making up the higher level entities—not these higher-level entities themselves. For the latter to be targets of the evolutionary process in their own right, it is *they* that need to be seen as differing in some traits, reproducing, and passing on their traits to their offspring. The fates of the particles making up the higher level entities, in and of themselves, need to be irrelevant for the analysis of the evolution of the assemblage (Godfrey-Smith, 2009, pp. 39–40; Okasha, 2006; Maynard Smith, 1987; Brandon, 1990).[15]

This leads to the next aspect of what it takes for something to start evolving on its own: *the conditions under which* the fates of the particles making up the higher-level entities are irrelevant for the evolution of the assemblage. Exactly what is needed to ensure that the particles of the assemblage are "held together" so strongly that the assemblage starts to evolve in its own right? It is not fully clear how to answer this question yet, but candidate suggestions include the existence of internal policing mechanisms or of an obligate division of labor among the particles making up the higher-order entity in question, such that these particles cannot persist or reproduce on their own (L. Buss, 1987; Maynard Smith & Szethmary, 1995; Michod, 1999; Okasha, 2006, chap. 8; Godfrey-Smith, 2009, pp. 121–118; Clarke, 2016).

Fortunately, for present purposes, the details of what it takes for an assemblage (or other entity) to evolve in its own right can be left open.[16] What matters here is just that there needs to be *something* that ensures that the particles are held together strongly enough to underwrite the assemblage evolving on its own. Further, it is important to note that, whatever this something turns out to be, it is likely to also depend on conditions in the wider environment of the assemblage in question (see also Godfrey-Smith, 2009, chap. 3). For example, how strong the internal policing mechanisms need to be likely depends on how much selection there is on cooperation. In cases where there is little selection on particle cooperation, much policing may be necessary to hold the particles together, whereas in cases with strong selection for particle cooperation, less policing may be necessary (L. Buss, 1987; Clarke, 2016; Maynard Smith & Szethmary, 1995; A. G. Hart, 2013). In short, to the extent that an assemblage lacks the mechanisms to hold its particles together (for internal or external reasons), to that extent it cannot be seen as the target of (selection-based) evolutionary processes (see also Godfrey-Smith, 2009). This point is key for what follows.

Given all of this, it becomes possible to use the model of the previous section to investigate the agential nature of firms from an evolutionary biological point of view. This can be done in two steps.

First, given the natural selection/market competition parallelism, the model can help determine when firms are genuine targets of (selection-based) evolutionary processes. If market competition is genuinely parallel to natural selection, then to the extent that firms are being selected by the market, they are genuine targets of (selection-based) evolutionary processes. This is precisely what the model can help address.

Specifically, the model shows that outlets really can, at least sometimes, be seen to be genuine targets of selection—and that is so even though they are composed of individual agents with their own, and to some extent outlet-opposing, agendas.[17] That is, the results of the model make clear that there are positive scenarios in which firms persist long enough to be selected and evolve in the appropriate, MLS2-based sense.[18] The entities—outlets—that survive and reproduce in the model can be seen to be collectives (made up of individual employees), but fully coherent evolutionary entities nonetheless.[19]

However, and equally importantly, the existence of the negative scenarios in the model further shows that, in many cases, outlets *lack* the needed internal coherence to be genuine targets of selection. In particular, it is possible that the individual agendas of the firms' employees are not balanced by sufficiently strong external pressures to maintain cohesive outlets so that these outlets are constantly pulled apart, causing their destruction and the formation of a new set of outlets or firms. As noted previously, there is little theoretical sense in speaking of outlets as targets of selection here—genuine *evolution* does not take place at all. In this way, the model shows that competitive markets can also make for environments in which the internal potential of firms to become targets of selection *cannot* be actualized (R. Nelson et al., 2018, pp. 131–132).

Crucially moreover, the model provides information about *when* each of these cases is likely to occur. In particular, as noted in the previous section, firms/outlets are more likely to be genuine targets of evolutionary selection (a) the lower the pressure towards "clock-punching" (parameter r)—which could be seen as related to the degree of "internal policing" in an outlet—and (b) the greater the selective pressures are towards cooperative interactions—in the form of parameters E_2 and E_3. In this way, the model can help determine *when* firms are genuine targets of (selection-based) evolutionary processes.

In turn, this matters for the second step of the present inquiry: the determination of when firms are agents of their own. In the background of this second step is the fact that the determination of the *evolutionary* nature of a firm has implications for the determination of its *agential* nature as well. There are two aspects to this implication (see also R. Nelson et al., 2018, p. 86).

On the one hand, if it turns out that—in situations generally conducive to underwriting the natural selection/market competition parallelism—firms are *not* targets of selection, then it is reasonable to infer that they are *not* agents of their own either. The reason for this is that if firms lack the necessary internal "connectedness" to be able to be maintained long enough to actually evolve, this suggests that they lack the internal stability to be stable economic agents as well. If firms are driven apart by the interests of their employees so quickly that the question of whether to expand in the market *never even arises*, then they cannot be seen as decision makers of their own.

Note that this is not to say that deciding whether to expand in the market is the only decision a firm has to make. Rather, the point is that the decision as to whether to expand in the market is a central decision that affects many other decisions as well—whether to invest in new capital, adjust production techniques, hire new workers, change the marketing, price the good in question, etc. It is not clear how a firm could make any of the latter decisions if it could not make the decision as to whether to expand in the market. Deciding whether to invest in new capital requires determining the growth trajectory of the firm, and much the same goes for the other decisions just mentioned. In turn, this suggests that firms that do not have the potential to make the decision as to whether to expand in the market cannot well be seen as agents of their own in any interesting sense of the word. They are just not in a position to make any of the major decisions a firm has to make. However, this is exactly true when it comes to the previous negative scenarios. In these cases, firms/outlets are torn apart so quickly that no firm/outlet is even in a position to decide whether to open offspring outlets and to expand in the market—and thus, any of the decisions connected to the latter. Hence, in these sorts of cases, firms are better seen as mere transactional spaces for their employees or as environments for behavioral routines to replicate in—i.e. in terms of one of the non-agential views.

On the other hand, if it turns out that (in situations generally conducive to underwriting the natural selection/market competition parallelism) firms *are* targets of selection, then this means it is at least *possible* that they are agents of their own. Note that this is not to say that the fact (when it is one) that firms are selected in an MLS2 sense *entails* an agency-based view of the firm. This is not so, as there is nothing incoherent about there being coherent transactional spaces (for example) that are selected as such. In this sense, therefore, the positive link between firms being targets of selection and their being agents of their own is weaker than the negative one. Still, it is worthwhile to note that the fact (when it is one) that firms are selected in an MLS2 sense underwrites the agency-based view of firms at least in the sense that it shows this view to be a genuine possibility that needs to be investigated further *in the situations in question*. This is worthwhile noting, as the determination of *the conditions under which* a holistic, collective view of agency is plausible is—as noted earlier—still in its infancy (Elster, 1982; Watkins, 1952; Hodgson, 2007a). Therefore, determining when it is the case that firms are genuine targets of selection is useful for determining when the agency-based view of the firm is even a possible contender to be taken seriously.

Therefore, the model used in this chapter does have some important implications for the agency-based view of the firm. Given the natural selection/market competition parallelism and the fact that there are connections between a firm being a target of a selective process and its being an agent of its own (though these connections are not one-to-one), the model can make clearer that:

(a) The agent-based view of the firm is less plausible when the destructive pressures on a firm (exerted by the individually beneficial actions of its employees) are not balanced by sufficiently strong market forces that reward highly profitable firms with increased presence in the market. This happens when employees quickly turn into clock-punchers, or if even highly cooperative outlets/firms are not strongly rewarded with increased representation in the market.

(b) The agent-based view of the firm is more plausible when the destructive pressures on a firm (exerted by the individually beneficial actions of its employees) are not balanced by sufficiently strong market forces that reward highly profitable firms with increased presence in the market. This happens when firm/outlet creation—not just firm/outlet survival—is sufficiently strongly related to the presence of high-effort employees to counterbalance the learning of these expectation-exceeders within outlets.

Two final points are important to note about these conclusions. First, they have some concrete implications for how to think about different firms in different sectors. In particular, these conclusions suggest that firms producing goods or services about which their employees are passionate are more likely to be agents of their own, as in those cases, there will be less of a tendency for employees to turn into clock-punchers (as the value of r will be lower there). This conclusion will be strengthened if there are strong pressures to expand in the market for those firms. Some examples of firms like these may be coffee shops or organic grocery stores. These firms often seem to have employees that remain committed for a long time (people are passionate about coffee and food) and where market presence depends strongly on the passion of the employees (people like their coffee shops and organic grocery stores small, welcoming, clean, well-cared for, expertly-run, etc.). By contrast, firms producing goods or services with which employees have a difficulty identifying—such as firms producing peripheral components of other goods (bottle tops for hand lotions, etc.)—or for which the pressure to expand in the market is not particularly strongly related to their cooperativeness—e.g. because production in the firm is heavily automated—are less well seen as agents of their own. Of course, further confirmation of this would require detailed empirical studies to see to what extent these firms and markets really do match the conditions of the positive and negative scenarios of the model used here.[20] The important point to emphasize is just that the present model should make these empirical investigations easier, as it can suggest what we ought to look for.

Second, it needs to be acknowledged that other models with different starting assumptions may well yield slightly different conclusions. However, this is not problematic, as it is a standard feature of model-based science: each model provides merely a glimpse or aspect of the phenomena under study (R. Nelson et al., 2018, pp. 130–133, 167; Wimsatt, 2007, chap. 6; Massimi, 2018). The key point to note here, though, is just that the present model allows us to learn something about the nature of firms using the natural selection/market competition parallelism. Since the model is based on core features of this parallelism, its conclusions should be taken seriously as such (whatever other conclusions we might derive elsewhere).[21]

VI. Conclusions

This chapter yields two sets of conclusions. On the one hand, there is a set of substantive conclusions concerning the agency-based view of the firm. This view is more plausible to the extent to which the market rewards intra-firm cooperation with firm reproduction, and less plausible to the extent in which it does not—modulated by the pressures on employees of firms to become cooperative, hard workers.

On the other hand, the arguments of the chapter also show more generally the promises and challenges of linking evolutionary biology and economics by means of a natural selection/market competition parallelism. In particular, I hope to have shown that it *can* indeed be reasonable to see natural selection and market competition as the same sort of process—but only sometimes. There are cases where competitive markets provide the conditions for genuine firm evolution to take place, in much the same way in which biological environments can provide the conditions for organisms to evolve. However, the two processes can also be quite different from each other. The conditions that underwrite a parallelism between these two processes are quite stringent. More importantly, the chapter also makes clear that appealing to the natural selection/market competition parallelism can be helpful for resolving open debates in economics (though of course it cannot resolve all the questions here). This last point is crucial, as, unlike in the case of the natural selection/economic choice parallelism, it is not nearly as clear what uses the appeal to a natural selection/market competition can have. By showing that such an appeal can help address the debate surrounding the nature of firms, therefore, the chapter aims to make clearer how this last question can be answered.

All in all, I hope to have further deepened the discussion of the benefits and challenges of the structural form of evolutionary economics. In particular, I hope to have shown that there is epistemic potential in this form of evolutionary economics. We really can learn something about economic questions by approaching them through structural parallels with evolutionary biological questions. However, I also hope to have shown that it cannot be presumed that this potential can always be actualized, and, even where it can be, it has to be done carefully: we need to pay close attention to the details of the specific case at hand. With this in mind, it is now useful to switch over to the evidential form of evolutionary economics.

Suggested Further Reading

R. Nelson and Winter (1982) is a classic study of the extent to which the spread behavioral routines through firms can be seen as an evolutionary process. R. Nelson et al. (2018) presents an overview of some recent work surrounding "firm evolution" (among other things). List and Pettit (2011) presents a wide ranging, philosophically grounded account of group agency. Godfrey-Smith (2009) and Maynard Smith and Szethmary (1995) are important discussions of the evolutionary pressures that lead to the emergence and maintenance of collective individuals.

Notes

1. For this reason, the analysis in Penrose (1952) is partly right and partly wrong. It is right in noting that the natural selection/market competition parallelism requires stringent conditions to be satisfied in order to get off the ground. However, it is wrong in that there is no reason to think that these conditions can never be satisfied.
2. A key historical application arguably lay in the question of whether we should expect most firms to actually be profit maximizers (Friedman, 1953). However, for reasons related to the ones just sketched, it quickly became clear that the answer to this question is complex, and that an evolutionary biological approach towards it is not obviously fruitful (see e.g. Alchian, 1950).
3. As will become clearer momentarily, there is no reason to assume that *all* firms are *always* one or the other. Indeed, determining *when* it is best to see *which* firms as agents of their own is the key issue to be addressed here.
4. R. Nelson and Winter (1982, pp. 124–128 and chap. 4) further argue that human agents can be subsumed under a similar model—they are embodiments of skills. If that is so, then this view would have the implication that there are no genuine agents at all that it is worth tracking in an economy. I will not consider this point further here, though.
5. So, Hodgson and Knudsen (2010, p. 173) note that "[t]he importance of . . . firm-specific capabilities and learning effects mean that the firm often has the necessary cohesion to qualify as an interactor," where an interactor is (following Hull, 1988) defined as an entity that "interacts as a cohesive whole with its environment in such a way that this interaction causes replication to be differential" (Hodgson & Knudsen, 2010, pp. 165, 167). However, the textual situation here is not entirely clear, in that Hodgson and Knudsen (2010) also seem to want to embrace the routine-based picture of the firm (see e.g. Hodgson & Knudsen, 2010, pp. 173–179). Assessing this is not so important here, though.
6. D. D. P. Johnson et al. (2013) consider a related set of issues within a somewhat similar theoretical framework. However, their explanandum is different—whether market competition leads to cooperation among individuals—and they also do not consider in detail the conditions required for a structural parallelism between natural selection and market competition to get off the ground.
7. This is thus akin to what is sometimes called a "Schumpeter Mark I" situation (R. Nelson et al., 2018, p. 194; R. Nelson & Winter, 1982).
8. This was done using NetLogo; the code is available at http://tinyurl.com/ofurolb.
9. These results have also been confirmed with a midlevel analysis that simultaneously varied r between 1, 4, and 7; E_1 between 0.01 to 0.41 in 0.1 increments; and E_2 and E_3 between 0.4 and 0.8 in 0.1 increments. The details of the results are available at http://tinyurl.com/pp4ec62.

10. This last fact should not be overemphasized—it is a straightforward consequence of the fact that this is a selection-based model with a monotonically increasing fitness function. At any rate, there can be long periods where several firms coexist in the market. Finally, as also noted earlier, the appearance of monopolies is neither an unexpected nor a problematic feature of cases of Schumpeterian competition (R. Nelson et al., 2018, pp. 107–118).

11. The number of expectation-exceeders there are in the initial pool of applicants (i.e. p_E) also matters, but it seems to primarily concern the length of time it takes before a firm establishes itself in the market.

12. Of course, this does not mean that these component parts cannot be targets of the evolutionary process in their own right as well.

13. Note also that this issue does not arise so much in the context of the natural selection/economic choice parallelism, as there (a) it is clear what is being selected on the economic side (actions or something related to this), and (b) it is irrelevant what is being selected on the biological side (genes, organisms, groups of organisms, etc.).

14. It is important to note that the second, MLS2-based case does not require that the fitness of an assemblage is completely unconnected to any of the features of its parts. The point is just that assemblage fitness is measured in terms of the expected reproductive success of assemblages—not of the individuals making up the assemblage. See also Okasha (2006).

15. This is also noteworthy due to the fact that prior investigations of this matter (such as that of Hodgson & Knudsen, 2010) have not considered the distinction between MLS1 and MLS2.

16. The better understood these conditions become, the more it becomes possible to use them to further the debate about the nature of the firm. It then becomes possible to lay out when we would expect firms to *turn into* agents of their own.

17. It is worthwhile to note that the model displays, in the first instance, *outlet* selection. *Firm* selection is merely a byproduct of outlet selection. This, though, does not alter anything of importance for the present discussion.

18. Here it is also important to recall (see also note 14) that MLS2 does not require group fitness to be completely independent of any property of the lower level (here, it is dependent on the number of expectation-exceeders in an outlet).

19. This conclusion is not altered by the fact that if one focuses on the evolution of the two *strategies* (expectation-exceeder/clock-puncher)—i.e. if one switches to considering the issue from the perspective of the routine-based view of the firm—the model appears to represent an MLS1-type scenario (within an outlet, the expectation-exceeder strategy does worse than the clock-puncher one, but assuming expectation-exceeder strategies cluster in some outlets, more new expectation-exceeder strategies will be created than clock-puncher ones *overall*). This change of perspective towards the evolution of strategies does not change the conclusion reached here, as the fact that this perspective is *available* does not mean that it should be *adopted*. The fate of the individual employees is of no direct consequence to the model at all. The total number of expectation-exceeders in the market can be taken to matter only to the extent that this influences the probability with which outlets survive and reproduce (in fact, individual employees need not be presumed to reproduce in this model at all). This point is further reinforced by the fact that this is an agent-based model, not a strategy-based one (see e.g. Alexander, 2007, for more on this distinction). Indeed, this sort of situation is exactly analogous to what is the case in many straightforwardly evolutionary biological cases. In the latter context, it is also often possible to keep track

of evolutionary changes in terms of changes in gene frequencies—however, this need not always be taken to be the best or only way of understanding the evolutionary mechanisms at play (Sober & Wilson, 1998; Okasha, 2006). For an MLS1-based look on a related set of issues, see also D. D. P. Johnson et al. (2013).

20. One confirmatory data point here is provided by the fact that some firms have found that increasing the number of full-time employees increases firm profitability—see Feintzeig (2016).

21. Of course, the model can also be used—perhaps in conjunction with other models—to investigate various other questions, e.g. concerning economic growth or innovation (see e.g. R. Nelson et al., 2018, for an overview of different such models).

4   Of Macaques and Men

The Comparative Approach Towards Economic Decision-Making (The Evidential Project I)

I. Introduction

One of the major areas of inquiry in economics is human decision-making, but this does not mean that economists are in agreement about how humans *in fact* make economic decisions.[1] Indeed (as also briefly noted in Chapter 2), there are many different theories of economic decision-making that are being defended, and it is not yet clear which of these (if any) should be seen to be the most widely confirmed one (see e.g. Binmore, 2007; Hausman, 2012; Gigerenzer & Selten, 2001; Kahneman, 2003; Glimcher et al., 2005; Loomes & Sugden, 1982). Given this ongoing lack of agreement on the best economic theory of choice, it becomes worthwhile to consider whether appealing to evolutionary biological considerations can be useful in this context (Hammerstein & Stevens, 2012). However, this appeal need not be seen to be restricted to the context of the structural form of evolutionary economics; the aim of this appeal could also just be evidential in nature. That is, the evolutionary biological perspective on economic decision-making might just be used to bring clarity as to which economic theory of decision-making is the most plausible one; no claim needs to be made—as was done in Chapter 2—that economic decision-making is *itself* an evolutionary process. Making clearer exactly how this is meant to work is the aim of this (and the next) chapter.

To do this, the focus here will be on comparative methods—a core part of the toolkit of contemporary evolutionary biology (Harvey & Pagel, 1991; Felsenstein, 2004).[2] (The next chapter considers a different, directly human-focused appeal to evolutionary biology to further the discussion surrounding economic decision-making.) Specifically, the aim in what follows is to analyze in more the detail the attempt by a number of researchers to use data on economic decision-making in non-human animals ("animals" in what follows) to advance our understanding of economic decision-making in humans (see e.g. Santos & Rosati, 2015; Kalenscher & van Wingerden, 2011; DeAngelo & Brosnan, 2013). As I show here, this appeal involves a number of complexities that need careful attention. In particular, it is important to get clearer on (a) exactly what the dispute surrounding human economic decision-making really comes down to, and (b) how data on animal economic decision-making can be related to this dispute.[3]

To make progress in answering these questions, I first (in section II) lay out the core types of economic theories of choice in more detail and defend a particular way of understanding the controversy surrounding them. Then, in section III, I develop a methodological framework for interpreting this data. In section IV, I apply this framework to some recent empirical work on animal decision-making. I present my conclusions in section V.

II. Economic Theories of Choice: Interpretations, Types, and Debates

What is the aim of a theory of human economic choice?[4] As mentioned in Chapters 1 and 2, there are three main interpretations of the aims of such a theory: a behavioral interpretation, a psychological/neuroscientific interpretation, and a normative interpretation. While a closer look at these interpretations was not necessary thus far, it is needed for the issues at stake in this chapter. Hence, I begin by considering these different interpretations in more detail. Once that is done, it is possible to lay out the different competing theories of choice in more detail. Finally, after that, I clarify the dispute surrounding these different theories.

1. Aims of an Economic Theory of Choice

The first possible aim of the economic study of choice is purely behavioral. In this interpretation, the goal of an economic theory of choice is setting out a (formal) framework that allows researchers to predict the choices—i.e. the behaviors resulting from decisions—that humans make in different circumstances (Friedman, 1953; Gul & Pesendorfer, 2008; P. A. Samuelson, 1938; Ross, 2005). In this approach, researchers are not concerned with developing the detailed psychological or cognitive neuroscientific mechanisms that bring about economic choice behavior—i.e. with *explaining* economic choices (at least on most accounts of explanation).[5] All that they are concerned with is providing an account that can predict and accommodate people's actual choice behaviors: what goes on in people's heads when they make decisions is not relevant to this way of understanding the economic study of choice.[6]

By contrast, it is precisely the psychological and neural mechanisms underlying choice behavior that are at the heart of the second possible interpretation of the aims of an economic theory of choice. To understand this better, it is useful to appeal to Marr's (1982) distinction among three levels of cognitive analysis: the computational level (the basic structure of the computational problem the cognitive system needs to solve), the algorithmic level (the precise way in which the computational problem is solved), and the implementational level (how the relevant computations are physically implemented).[7] Given this, the core thought behind the psychological/cognitive neuroscientific reading of economic theories of choice is to specify not just the computational problem the agent needs to solve, but also *at least* the algorithmic underpinnings of its choice behavior. It tries to answer the question of how agents actually make economic decisions and what sort of mental computations they engage in when they make decisions (Kahneman & Thaler, 1991; Thaler, 2000; Glimcher et al., 2005; Thaler, 1980; Rabin, 1998; Hutchinson & Gigerenzer, 2005, pp. 105, 110–111, 119). A recent development of this sort of approach towards economic decision-making goes even further by considering the implementational level as well. Specifically, neuroeconomists have advertised the study of the neural structures underlying economic decision-making as being both inherently important and informative of the psychological mechanisms underlying economic choice behavior (Glimcher et al., 2005; Fehr & Camerer, 2007; Camerer, 2007).

At this point, it needs to be noted that, while there is no question that there are some important differences between these two interpretations, a closer look at them also reveals some important communalities. In particular, in order to avoid known problems for the behavioral interpretation, some commitment to psychological mechanisms needs to be made there, too. For example, it is now widely accepted that economists—e.g. when it comes to strategic choice situations—need to be able to assess the choices that agents *would* make, *were* they in a certain situation (Hausman, 2012, 2008). In turn, this implies that the goal of any economic theory of choice needs to go beyond describing the choices that agents actually make, and at least commit to *dispositions* to choose in certain ways (see e.g. Binmore, 1998, 2007; Guala, 2012; Wade Hands, 2012). However, such a commitment to *dispositions to choose* at least indirectly makes some claims about the psychological structures underlying choice behaviors—it is just that these structures are not described in any kind of detail. In this way, it is best to see the behavioral and the psychological/ neurophysiological interpretations as differing in degree, rather than in kind. However, this is of course not to deny that these two interpretations are importantly different. After all, the behavioral interpretation seeks, as much as possible, to *avoid* committing to the *exact* psychological or neurophysiological mechanisms underlying choice behaviors, whereas identifying these mechanisms is the very heart of the psychological/ neurophysiological interpretation.

Both of these aims differ quite strongly from those of the third, normative interpretation of the economic theory of choice. According to this interpretation, the goal of such a theory is neither the specification of the choice behaviors actual agents engage in, nor in specifying the psychological or neural mechanisms underlying their choice behaviors; rather, the goal is the specification of the choice behaviors or decision-making procedures agents need to engage in *so as to be rational*.[8] In this reading, then, what people actually do or how they actually think is not so important, what matters is what they *should* do or think. Note also that, as such, in this reading, economic theories of choice neither explain nor predict actual choice behavior—they only specify rational choice behavior. Of course, on the assumption that people are actually rational, this specification may have predictive (and perhaps even explanatory) implications. However, these implications are not part of the theory, but result from an extraneous assumption—the fact that (the relevant) agents in fact are rational—that needs to be independently supported. In a normative reading, economic theories of choice at most try to explain and predict choices *via* the assumption that people are (largely or at least sometimes) rational, but the latter is an auxiliary assumption that can be questioned without calling into doubt the fundamental claims of an economic theory of choice (Lyons et al., 1992; Joyce, 1999).

However, for present purposes, this normative reading will not be central. This is not because there is anything deeply wrong with seeing economic theories of choice as providing a logic of rational choice. Rather, the focus on the first two (positivistic) readings is due to the fact that these readings are most conducive to an (evidential) evolutionary theoretic approach. These readings do not need to commit to some controversial normative auxiliary principles, such as a strong "ought implies can" principle (according to which a normative standard is open to criticism if it can be shown that the relevant economic agents have not evolved the kinds of abilities to satisfy this standard) or a denial of the naturalistic fallacy (according to which claims about what has evolved entail claims about what rationally ought to be the case).[9] While there may well be ways to justify these further normative principles, doing this will require significantly further argumentative work. It is also important that this argumentative work needs to come *on top of the* establishment of linkages between evolutionary biology and positive readings of the economic theory of choice. It *first* needs to be shown what implications evolutionary biological considerations have for how people do choose; only then we can ask what this implies about how they ought to choose (see also Santos & Rosati, 2015; Kalenscher & van Wingerden, 2011).

For these reasons, the focus here will be squarely on the positive readings of the economic theory of choice (for some discussion of evolutionary arguments for the normative reading, see e.g. Gigerenzer & Selten, 2001; Gigerenzer, 2008; Hagen et al., 2012; Santos & Rosati, 2015). That is, the question in what follows is what studies of animal economic decision-making can tell us about the choices that people *actually* make and/or the psychological or neurophysiological mechanisms *actually* underlying these choices.[10] Keeping this non-normative focus of the discussion in mind is very important for what follows.

2. Four Types of Economic Theories of Choice

There is a vast number of different economic theories of choice that have been offered. Fortunately, for present purposes, a complete overview of them is not needed, as it is possible to see these many different theories as clustering around a small number of poles. Four of these poles are especially important: (a) classical rational choice theory, (b) prospect theory, (c) regret theory, and (d) the theory of simple heuristics. While this still leaves out of consideration some important alternative theories—such as Busemeyer & Townsend's (1993) decision field theory and various theories dealing with choice under conditions of (radical) uncertainty (see e.g. Ellsberg, 1961; Voorhoeve et al., 2016; Wakker, 2010)—this is not greatly problematic here. This is so for two reasons.

First, and most importantly, these four theories represent four major ways of understanding the psychology of economic decision-making: as based on reasoning about outcomes and probabilities, as based on reasoning about prospects and decision weights, as based on reasoning about counterfactual possibilities, and as based on many domain-specific simple heuristics. These four types of theories thus mark four different poles around which decision-making could be oriented. Second, focusing on these found theories is not problematic due to the fact that there is either no data from animal decision-making that clearly bear on the theories that have been left out, or that the same issues that are being raised for the four types of theories (a)–(d) also apply to the ones that have been left out (as in the case of Ellsberg's paradox for example: Hayden et al., 2010; Rosati & Hare, 2011). Leaving these other theories out of consideration thus makes the discussion here more manageable without affecting the scope of its conclusions drastically. Consider the four different types of theories (a)–(d) in more detail.

a. Classical Rational Choice Theory

At the heart of the set of classical rational choice theories (RCT in what follows) is the idea that economic agents make decisions by assessing which action leads to the most desirable outcome on average (Jeffrey, 1983; Savage, 1954; Hausman, 2012; Resnik, 1987). In a bit more detail (expanding on some of the remarks of Chapter 2), this theory works as follows.

Agents are assumed to partition the world into a set S of n exhaustive and mutually exclusive states of affairs. The agent then considers what she believes to be the set A of m different actions open to her, as well as what she believes are the $(A \times S)$ consequences of these different actions in the different states of the world. Each of the consequences in $(A \times S)$ is evaluated as to its desirability by assigning it some utility value. Each of the actions in the set A of feasible actions is then assigned an expected utility value—i.e. the probabilistically grounded (typically arithmetic) average over the utility values of the different possible consequences of the action in the different states of the world. Formally:

(1) $EU(A_k) = \sum_{i=1}^{n} P(S_i) u(A_k \& S_i)$

where $P(x)$ is a probability function ranging over S and $u(x)$ is a real-valued and only minimally cardinal utility function ranging over $(A \times S)$.[11] Finally, agents are assumed to choose the action whose value of $EU(x)$ is greater than that of all of the alternatives. Three points are important to note about RCT in the present context.

First, on this type of theory, what the agent is assumed to evaluate are *consequences*—i.e. the outcomes of actions in different states of world. Importantly, these evaluations are independent of how the relevant consequences compare to the status quo, or what other consequences the agent could have brought about by making other choices. The agent is assumed to concentrate on just the states of the world that result from its behavior.

Second, in evaluating actions, the agent considers the (subjective) *probabilities* of the different states of the world occurring.[12] In turn, this implies that shifts in the relevant probability function that affect all states of the world equally do not alter the outcome of which actions the theory expects to be chosen.

Finally, the manner in which the agent is assumed to evaluate actions is in terms of a *maximization* process. As noted in Chapter 2, in RCT agents are assumed to look for the action that is associated with the highest expected utility—not one that is associated with a "sufficiently high" such value (or something similar).

b. Prospect Theory

In a number of publications over several decades, Daniel Kahneman, Amos Tversky, and their colleagues have championed the idea that much of human decision-making—indeed, much human cognition more generally—relies on a set of "heuristics and biases." In particular, they have argued that, instead of maximizing their expected utility as suggested by RCT, people make decisions in a much more situation- and frame-dependent manner that relies on a number of cognitive shortcuts. While there are, again, many different ways of spelling out their research program (see e.g. Barberis, 2013; Kőszegi & Rabin, 2006; Tversky & Kahneman, 1992; Kahneman et al., 1982), for present purposes, it is sufficient to consider the outlines of their classic prospect theory (PT in what follows) (see e.g. Kahneman & Tversky, 1979; Barberis, 2013).[13]

According to PT, people make decisions in some ways like RCT says they do, but, instead of evaluating consequences (as defined previously), people are assumed to evaluate *prospects*—changes to their status quo. Also, instead of considering the (subjective) probabilities associated with different states of the world, people consider "*decision weights*": transformed probabilities that give more weights to the tails of the probability distribution. Slightly more formally, this idea can be expressed as follows.

People assess prospects with a value function $v_C(x)$ that ranges over the prospects the agent believes to be available to her in her situation C. This value function is furthermore assumed to be concave in gains and convex in losses. People are risk-averse in gains (they prefer a certain gain of G to a gamble with the same expected gain of G), but risk-loving in losses (they prefer a gamble with an expected loss of L to a certain loss of L). Figure 4.1 makes this clearer.

Further, people assess risky prospects with a decision weight function $p(x)$ ranging over the possible states of the world S, which is steeper near the tails, as in Figure 4.2.[14]

Finally, they combine $p(x)$ and $v(x)$ by choosing the action A_k that maximizes:

(2) $EP(A_k) = \sum_{i=1}^{n} \pi(S_i) v_C(A_k \& S_i)$

For present purposes, four points are important to note about PT.

First and most importantly, according to PT, people evaluate *prospects*—gains and losses—not outcomes (i.e. consequences). This is an important

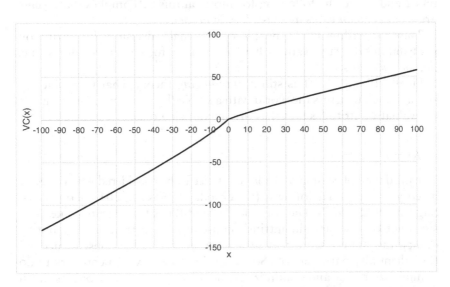

Figure 4.1 A plot of value function $V_C(x) = -2.25(-x)^{0.88}$ if $x < 0$; $V_C(x) = x^{0.88}$ if $x > 0$

Source: See Tversky and Kahneman (1992) for a related figure

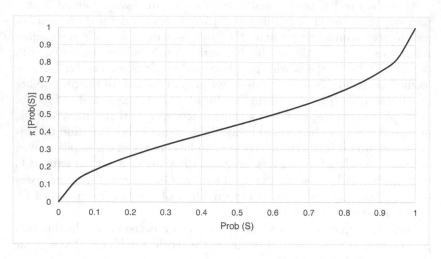

Figure 4.2 A plot of $\pi[P(S)] = P(S)^{0.65}/ [P(S)^{0.65} + (1 - P(S))^{0.65}]^{1/0.65}$

Source: See Tversky and Kahneman (1992) for a related figure

departure from RCT, as it makes PT more situation-dependent than the latter; I return to this point momentarily.

Second, there is a specific set of attitudes towards risk built into PT. Unlike RCT, which, depending on the shape of the utility function, can make room for many different kinds of attitudes towards risk (see Chapters 2 and 5 and the following for more on this), PT implies that agents are *risk-averse in gains and risk-loving in losses.*

Third, according to PT, people do not make decisions by relying on probabilities directly. Rather, they rely on transformed probabilities—i.e. *decision weights.*

Finally, like RCT, PT is still a *maximizing* theory. That is, people are still assumed to assess the world with a stable (but context-dependent) set of evaluative attitudes that they are maximizing over.

c. Regret Theory

The third type of economic theory of choice to be considered here is Regret Theory (RT in what follows) (Bleichrodt & Wakker, 2015; Loomes & Sugden, 1982, 1987). The core idea behind RT is that agents make decisions not by assessing the intrinsic consequences of their actions, but by comparing the consequences of a given action to the consequences of the salient alternative action. So, according to RT, what people try to do is minimize regret and maximize rejoicing: they want to avoid, as much as that is possible, ending up in a situation where they wish they had done something different from what they actually did. Again, this can be spelled out in a number of different ways (see e.g. Bleichrodt & Wakker, 2015), but the basic idea is this.[15]

Agents are assumed to evaluate the world in terms of a function $Q(x)$ which ranges over utility differences among pairs of actions. Specifically, $Q(x) = f(u(A_k) - u(A_l))$, where A_l is the focal alternative action.[16] Furthermore $Q(x)$ is typically assumed to be symmetric, in the sense that $Q(-x) = -Q(x)$, and to have various further features (such as being convex for "rejoicings," i.e. positive values of $Q(x)$ and concave for regrets, i.e. negative values of $Q(x)$). Finally, RT assumes that agents make decisions by maximizing:

$$(3) \quad ER(A_k, A_1) = \sum_{i=1}^{n} P(S_i) \, Q[u(A_k \& S_i) - u(A_l \& S_i)]$$

Three further points are important to note about this theory.

First, this theory can easily allow for *intransitive choices*. In particular, it is entirely possible for $ER(x, y) > ER(y, x)$, $ER(y, z) > ER(z, y)$, and yet $ER(z, x) > ER(x, z)$. This is quite intuitive, as $ER(.)$ values are inherently comparative; hence, changing the comparison class of a given action (whether x is compared to y or z) can change the expected regret value of the action as well.

Second, it is important to note that while both RT and PT are relational, the way they are *relational* is quite different. In particular, RT compares the realized consequences with unrealized, merely counterfactual consequences. By contrast, PT compares realized consequences with the realized status quo (i.e. the agent's initial situation).

Finally, like both RCT and PT, RT is still an inherently *maximizing theory*. While what is being maximized here is different, the basic idea that agents maximize over an evaluative space is the same.

d. The Theory of Simple Heuristics

Gigerenzer et al. have famously have argued that the maximizing nature of the previous three theories of choice is deeply mistaken (see e.g. Gigerenzer & Selten, 2001; P. Todd & Gigerenzer, 2007). They think that economic decision-making is best seen as based on a large number of "simple heuristics." While there are again many different ways of spelling out this idea, the main gist behind the theory of simple heuristics (SH in what follows) can be laid out like this.[17]

First, simple heuristics are easy to apply: they are structured so that their computational requirements are low, they do not require much information to function, and they can be applied quickly (Gigerenzer & Selten, 2001; P. M. Todd, 2001; Gigerenzer, 2007).[18] Second and relatedly, simple heuristics do not aim for optimality in the behavioral responses they yield. More specifically, simple heuristics are "satisficing" in that they lead to outcomes that, while acceptable, do not directly aim at the optimal decision.[19] Third and finally, agents relying on simple heuristics will rely on many different decision rules in many different circumstances. To function adequately, a simple heuristic needs to be applied to the appropriate situation (Gigerenzer & Selten, 2001, p. 107; Kruglanski & Gigerenzer, 2011; P. Todd & Gigerenzer, 2007; Hutchinson & Gigerenzer, 2005). Indeed, this last point is the flipside of the first two. The reason why simple heuristics allow for adequate decision-making that is at the same time fast and frugal is that they employ features of the decision situation (i.e. the agent's environment). However, this then implies that different decision situations—i.e. different environments—are likely to require different simple heuristics to allow for adequate decision-making (P. Todd & Gigerenzer, 2007). An example might make this clearer.

Assume an agent needs to decide which of two decision options to pursue. Assume further that, in this context, the familiarity of a decision option correlates (not necessarily perfectly) with its "choiceworthiness" (the extent to which it yields the optimal outcome). If so, then an agent can make adequate decisions quickly and frugally just by picking the decision options with which they are (most) familiar (P. Todd & Gigerenzer, 2007). However, this simple heuristic will not yield satisfactory outcome in cases where familiarity does not correlate (sufficiently well) with "choiceworthiness." In the latter case, the agent may need to rely on a different heuristic, such as "Take the Best," where the agent only evaluates the options according to one decision dimension (e.g. price) at a time, until a discrimination has been reached (Hutchinson & Gigerenzer, 2005).

In this way, decision-making based on simple heuristics can be seen to be exactly the opposite of decision-making as envisioned by traditional accounts: it is *domain-specific*, *pluralistic* and *satisficing*. Agents are not assumed to make all (economic) decisions in the same way, but rely on a battery of different decision rules, each of which is easy to apply and does not aim at finding the highest point in the preference value landscape, but merely to make for situationally satisfactory decision outcomes.

Having thus laid out the major types of economic theories of choice, it now needs to be noted that there is still much controversy over which of these four theories is the most plausible one. The next sub-section considers the nature of this controversy in more detail.

3. Controversies in Understanding Economic Choice

A traditional way of framing the controversy surrounding these different theories is in terms of departures from economic rationality. RCT is seen as laying down the economic decisions humans ought to make, and PT, RT, and SH as ways of accommodating the decision-making biases and quirks people actually display (see e.g. Kalenscher & van Wingerden, 2011; Gigerenzer & Selten, 2001; Santos & Rosati, 2015; V. Smith, 2007;

Kenrick et al., 2009; Boudry et al., 2015; Polonioli, 2015). However, as should be clear given the previous remarks concerning the aims of an economic theory of choice, this is not, in fact, a good way of understanding the conflict among RCT, PT, RT, and SH.

On the one hand, this conflates the different readings of the theories of choice—it combines a normative understanding of RCT with positive understandings of the other theories. This, though, leads to there not being a clear conflict between these different theories; if RCT just makes for a rational standard of decision-making, then the fact that people do not obey this standard raises no more problems than the fact that people do not always act morally or reason in line with accepted standards of theoretical rationality (such as principles of first-order logic).[20] On the other hand, and more importantly, this "deviations from homo economicus" reading of the conflict fails to do justice to the work that is being done with RCT, PT, RT, and SH. It fails to do justice to the work being done with RCT, as this theory has been and continues to be put forward as a positive theory—a point to which I return momentarily (Glimcher et al., 2005; Binmore, 1998, 2007).[21] It fails to do justice to PT, RT, and SH, as these theories have been and are put forward as genuine theories of economic decision-making in their own right, and not merely as deviations from RCT (Barberis, 2013; Bleichrodt & Wakker, 2015; Hutchinson & Gigerenzer, 2005; V. L. Smith, 2005).

Given this, a much better way of seeing the conflict among these four theories is by taking all of them seriously as positive theories of choice. That is, it is best to see the conflict here as coming down to the question of which of these theories is to be seen as the most plausible one for explaining and predicting actual economic decision-making, without any one of them being given some sort of default status. The reason for why there is a conflict between them then is precisely the fact that all of these theories can make *some sense* of *all* of the data, and that for each theory, there is a set of data for which it is *especially* compelling (relative to its competitors)—each theory *predicts* a set of findings that the others can only *accommodate*. (I return to this distinction between prediction and accommodation momentarily.) Overall, this leads to a case where no theory can be ruled out by the current data, and no theory can be clearly shown to be superior to all others when all the current data are considered.[22] Four sets of examples make this clearer.[23]

First, there are several situations in which RCT has been shown to make highly accurate predictions about how people choose. In particular, in many strategic interactions in which agents are given sufficient time to learn from their environment, their behavior is well accounted for by seeing it as maximizing a probability-weighted outcome-focused evaluative measure (this is a classic point in the literature; see e.g. Binmore, 2007; Glimcher et al., 2005; Schumpeter, 1959, p. 80). For example, Glimcher et al. (2005) have found that in the so-called "shirking game" (see

Table 4.1 The shirking game; the shirking rate at the Nash equilibrium is $s = i/w$

Employee/Employer	Inspect	Don't Inspect
Work	w-c/v-i-w	w-c/v-w
Shirk	0/-i	w/-w

Table 4.1), people match the predictions of RCT extremely well. (They only diverge from the equilibrium shirking rate of s if s is less than 0.3, in which case they "over-shirk," but this may be due to their inability to assess the relevant probabilities accurately: Glimcher et al., 2005).

However, this does not mean that PT, RT, and SH cannot make sense of these findings. In the first place, the way in which prospects are individuated in PT is left open, so that it is entirely possible to argue that, in the present cases, prospects and outcomes match very closely. Similarly, RT can allow that, with significant learning periods, people focus more on the actual outcomes of their actions and less on the regret/rejoicing that accompanies these outcomes (Bleichrodt & Wakker, 2015). Finally, defenders of SH can note that, when people have enough time to learn the structure of the decision situation they are in, they will determine which heuristics match the optimal outcomes particularly well (P. Todd & Gigerenzer, 2007; Kruglanski & Gigerenzer, 2011).

So, the conclusion here is not that the findings from many strategic interactions *rule out* PT, RT, or SH. Rather, the point is that these findings evidentially *favor* RCT over the last three theories. At the heart of this evidential favoring is the fact that only RCT *predicts* these findings, whereas PT, RT, and SH merely *accommodate* them. There is much that can be said about this notion of evidential favoring through predictive rather than accommodative fit—including how it could be spelled formally (Sober, 2008; Hitchcock & Sober, 2004). However, all that matters here is that it is widely accepted that, ceteris paribus, *predicted fit* to the data has more evidential weight than merely *accommodated fit*: the latter may be a sign that the data "overfits" the data, and thus mistakes the noise in the data for the signal (Hitchcock & Sober, 2004). Hence, findings like those of Glimcher et al. (2005) can be seen to constitute *evidence* for RCT over PT, RT, and SH—nothing more, but also nothing less.[24]

Second, consider the set of findings that people display: (a) the endowment effect (the finding that people need to be compensated more to sell an item that they own than what they are willing to pay to obtain that item), (b) framing effects (where people make decisions differently depending on whether they are framed as gains or losses), and (c) risk aversion in gains/risk love in losses (where people are more willing to gamble to avoid a loss than to obtain a gain). All of these findings are directly predicted by the theory (Kahneman & Tversky, 1979; Kahneman et al., 1982; Kahneman & Thaler, 1991; Barberis, 2013).[25] However, this does not mean that the other theories cannot account for this fact (see also Barberis, 2013).

So, RT has an easy time accounting for these findings, as the Q(x) function at its heart can be (and often is) made to have the kind of form that would lead to the Endowment Effect and risk aversion in gains/risk love in losses (Loomes & Sugden, 1987, 1982). The same goes for SH, which, in the worst case, can just postulate an "ownership heuristic." In fact, even RCT can account for these findings. On the one hand, RCT leaves it open how the consequences it ranges over are individuated. This means that it is able to define these consequences very finely, so that the state of the world in which the agent's wealth is increased by $100 is seen as different depending on how this increase came about (through a gain or a loss).[26] On the other hand, RCT leaves the shape of the agent's utility function open. This means that this theory is consistent with highly non-linear and irregular utility functions—and thus with the previously mentioned findings of the Endowment Effect and risk aversion/risk love. Note that this is not to say that these other theories have as easy of a time dealing with these findings as PT. As before, only PT straightforwardly *predicts* these findings; the other theories can merely *accommodate* them (see e.g. Rabin & Thaler, 2001). In turn, this means, again, that the existence of these kinds of effects is *evidence* of PT over RT, SH, and RCT— but doesn't rule out the latter.

The third set of findings to consider here concerns the fact that many people make intransitive choices: in a choice between A and B, they choose A; in choice between B and C, they choose B; but in a choice between A and C, they choose C. This sort of choice pattern has been found in many different circumstances, and thus appears very widespread (though less than initially thought: Bleichrodt & Wakker, 2015; for further discussion, see Kalenscher et al., 2010; Tsai & Bockenholt, 2006; Sopher & Gigliotti, 1993; Loomes et al., 1991). Now, as noted previously, RT can make sense of this very easily. Given that it sees decision-making as depending not just on the actual outcomes of the action in question, but also on the outcomes of the actions it is compared with, intransitivity can be predicted to occur (Loomes & Sugden, 1987, 1982; Loomes et al., 1991). Once again, though, this does not mean that PT, RCT, or SH cannot make sense of intransitive choices.

So, PT can, at least sometimes, explain choice intransitivity by noting that the agent's status quo has changed: from its initial standpoint, A might have been a more valuable prospect than B; then, from the standpoint of having just chosen A, B might have been a more valuable prospect than C; and then, finally, from its standpoint of just having chosen B, C might have been a more valuable prospect than A. As such this may be entirely coherent—it depends on how the relevant prospects are individuated.[27] As far as RCT is concerned, there are two main ways to explain intransitive choices (Broome, 1991b; J. G. Johnson & Busemeyer, 2005). First, it may be that the relevant agent has changed her mind between the first and the third choice: so while the agent first set EU(A) > EU(B) > EU(C), they then—perhaps even as a result of being confronted with the choice between A and C (Cubitt et al., 2004; Sen, 1997)—change their mind and set EU'(C) > EU'(A).[28] Second, as in the previous example, it may be that the outcome space needs to be re-individuated here. In particular, as far as the agent in question is concerned, it may be that the A in the first choice and the A in the third choice are actually different actions. If so, then the agent really takes it to be the case that EU(A1) > EU(B) > EU(C) > EU(A2), which is entirely transitive. Finally, due to the fact that different decision situations likely lead to the use of different simple heuristics, SH is also consistent with the fact that people's choices are often intransitive. Indeed, transitivity in choices is not necessarily to be expected on SH (though neither is its opposite). As before, it needs to be noted that this is not to say that PT, RCT, or SH can make sense of intransitive choice as easily as RT can; they can *accommodate* it, whereas the former *predicts* it. Hence, choice intransitivities can be seen as evidence for RT over RCT, PT, and SH.

Finally, for the last type of evidence, consider the fact that, out of a range of options, people tend to choose the one that they are familiar with (Henrich & McElreath, 2007; Goldstein & Gigerenzer, 2002). This is something that is straightforwardly predicted by the "recognition heuristic" of SH. In choice situations where only one option is known, people use the recognition heuristic to choose that one (and often pick appropriately). Similarly, consider the fact that people often seem to make decisions using a lexicographic hierarchy of simple cues—as soon as a cue is found that discriminates among the options, they stop (Dhami, 2003; Broder, 2000). However, this does not mean that the other theories cannot make sense of this finding.

So, defenders of RCT can note that people place a lower utility on unfamiliar items, or that people's utility ranges over both the items in question and decision-making time, so that expected utility maximization overall can require abandoning further deliberation once a reasonably good estimate of the expected utility value of the relevant options has been reached. Much the same is true for defenders of PT or RT—who would just replace "expected utility" in the previous sentence with "weighted prospect" or "expected rejoicing." However, it seems once again plausible that this sort of finding makes for evidence for SH over RCT, PT, or RT. While it may be possible to somehow account for these findings using these other theories, it is most straightforwardly predicted by SH-based decision-making.

All in all, this thus shows that each theory has a domain in which it does predictively better than its competitors—though each theory can, somehow, account for all the findings.[29] Of course, as noted earlier, this assessment is based on only a small sample of findings concerning human economic decision-making—no attempt at a systematic review of the literature was made (which would be very difficult, given the size of this literature). Still, what this summary brings out is the *nature* of the controversy surrounding the key economic theories of choice. At the heart of this controversy is the fact that the currently available data about the way humans make economic decisions do not clearly favor one of the "poles" (outcome-focused/prospect-focused/counterfactual-focused/process-focused) over the others.

It is furthermore important to note that, in principle, there are two different ways in which this ambiguity could be resolved. These two different ways turn on two possibilities about the relationship between the different economic theories of choice (RCT, PT, RT, and SH), and the nature of the human economic decision-making mechanism.

On the one hand, one could see the different theories as different and competing descriptions of *one* unitary human economic decision-making mechanism. That is, perhaps humans make all economic decisions in the same way, and what we are trying to determine is exactly *how* they do so. In this view, each of the previously mentioned theories proposes a different way of characterizing this unitary mechanism underlying human economic decision-making. (Of course, it is in the nature of SH that it sees this mechanism as constituted by a set of simple heuristics.) Our question is which of them is the best such characterization.

On the other hand, one could see the different theories described here as different but non-competing descriptions of different *aspects* of the same, pluralistically structured human economic decision-making mechanism.[30] That is, perhaps humans make different economic decisions in a different manner: they employ different decision-making mechanisms in different circumstances. In this view, each of the theories sets out a different part of this pluralistic mechanism. Our question is which of them is triggered in which circumstances—when do humans rely on an RCT-like decision-making mechanism, when on a PT-like one, and so on (see also Barberis, 2013, pp. 179, 192).[31] While we have a sense of some of the cases in which RCT (for example) is particularly predictively successful—involving significant learning periods, for example—we do not know what the range is of these cases exactly, nor do we know *why* RCT-based decision-making is particularly pronounced *in these cases*. The same goes for PT, RT, and SH; we may know of some of the situations in which each of these theories is evidentially favored, but we do not know the extent of these situations or what they have in common that leads them to trigger the relevant decision-making mechanisms. In what follows, I shall refer to this issue as the determination of the *domain of application* of the different decision-making mechanisms.

With this in mind, it now becomes possible to make more precise the nature of the conflict surrounding the different theories of human economic decision-making. This conflict can be summarized by two related sets of questions:

[1] Is there one unitary human economic decision-making mechanism, or is the latter subserved by a disunified collection of different economic decision-making mechanisms?

[2] [a] If there is only one unitary human economic decision-making mechanism, what is it like—i.e. which of these theories best characterizes it?

[b] If human economic decision-making mechanism is subserved by a disunified collection of different economic decision-making mechanisms, what are the domains of application of its components—i.e. which decision situations are best characterized by which of the theories?

In what follows, I take answering these two questions to be the main challenge in understanding human economic decision-making (see also Santos & Rosati, 2015, pp. 3, 19 for hints of a similar view about the nature of the conflict here.).

A presupposition in the background of questions [1] and [2][a]/[b] is that it is not the case that *none* of these theories (RCT, PT, RT, or SH) is to be seen as compelling, and that a search needs to be started for a new and radically different kind of economic theory of choice. This presupposition is plausible, since the four types of theories here represent the four key poles around which decision-making could turn (outcome-focused, prospect-focused, counterfactual-focused, and process-focused), and since they all make some confirmed, surprising predictions. In turn, this suggests that it is not warranted to think that they are *all* on the completely wrong path: the fact that they all have a (more or less) easy time accommodating the data, in combination with the fact that they all make surprising predictions of their own, suggests that they should be seen to get at *something* concerning the processing underlying and leading up to economic choices.

Now, as noted earlier, the main reason why answering questions [1] and [2][a]/[b] is such a challenge is that, as it stands, the data from human economic decision-making are too ambiguous. As of yet, each of the many different theories available has some empirical successes—i.e. makes predictions that can only be accommodated by the other theories—and we do not yet have a good sense of whether one of these theories will ultimately turn out to be more compelling than the others, or whether they are all right in their own domains—of which, though, we do not yet have a good grasp either (see also Harless & Camerer, 1994). To make progress here, then, more data are needed. These new data could show that one of these theories is, after all, empirically superior to the rest—by being the most predictively accurate overall (Hitchcock & Sober, 2004)—or they could deepen the view that economic decision-making is pluralistic if all theories continue to be equally predictively accurate. If the latter, the new data may also be able to show how the different parts of human economic decision-making interact.[32]

Where, though, are these new data going to come from? One obvious answer is from future studies of human economic decision-making. However, there is also another (complementary) answer that has recently been advertised (Santos & Rosati, 2015; Kalenscher & van Wingerden, 2011; Hammerstein & Stevens, 2012), and which does not require us to await the (perhaps quite far off) results of future work: namely, from studies on animal economic decision-making. It is this answer that is the focus of the rest of this chapter.

III. A Framework for the Appeal to Animal Decision-making

Why would appealing to how animals make decisions be interesting if we are concerned with how humans make economic decisions? The obvious answer is that humans are biological organisms, too. In turn, this makes what we know about economic decision-making in other organisms evidentially relevant to the investigation of human economic decision-making. In particular, it is a standard assumption of phylogenetic approaches towards evolutionary biological phenomena that related organisms share traits—and the more so, the more closely related they are (Santos & Rosati, 2015; Hutchinson & Gigerenzer, 2005; Kalenscher & van Wingerden, 2011; DeAngelo & Brosnan, 2013; Harvey & Pagel, 1991; Felsenstein, 2004). So, if we can determine that a type of organism O_1 makes decisions in a certain way, then this provides evidence for the fact that a related type of organism O_2 also makes decisions in this way; given their common ancestry, we can expect these two types of organism to make decisions in similar ways. Of course, as will also be made clearer later on, this is not guaranteed—organisms can evolve away from their common ancestor. Still, the fact that O_1 makes decisions in a certain way is *a reason* for thinking that O_2 does so too. Short of countervailing considerations, this reason can and needs to be taken seriously in its own right (Harvey & Pagel, 1991; Felsenstein, 2004). In this manner, the consideration of animal economic decision-making can broaden the set of data we can theorize over—and thus hopefully help us make progress in resolving questions [1]–[2][a]/[b].

However, before this broadening of data under consideration can begin, a preliminary worry needs to be addressed. In particular, one may be concerned that human economic decision-making is uniquely human—i.e. that it does not have any interesting counterparts among animals. So, just like it is widely accepted that humans are unique in communicating in a compositional, productive, and systematic manner—i.e. linguistically—it may be that humans are unique in making economic decisions. This may be underwritten by the thought that genuine economic decision-making depends on a suite of distinctively human mental representations including notions of self, property, value, cost, and price (Steele, 1988; Beggan, 1992; see also C. Taylor, 1971). If this is right, then it might be thought that there is little reason to look towards animals in studying human economic decision-making if the latter is simply a distinctively human trait. However, for three reasons, this human-restricted view of economic decision-making ought to be resisted.

First, the dependence on specifically human mental competences is not obviously part of these theories concerning human economic decision-making. There is nothing in the nature of RCT, PT, RT, or SH that suggests that they depend on rich concepts of self, property, value, cost, price, or the like. Of course, it is possible that this dependence is masked—a point to which I return momentarily—but, at least on the face of it, this dependence on (alleged) specifically human concepts is not visible in the structure of these theories.

Second, comparative studies have been shown to be useful even for seemingly human-specific traits like natural languages (see e.g. Sterelny, 2012c; Scott-Phillips, 2015; Jackendoff, 1999). That is, while it is widely accepted that humans are unique in communicating in a compositional, productive, and systematic manner, it is also widely accepted that many other animals communicate in complex ways as well (see e.g. Russon & Andrews, 2010; Scott-Phillips, 2015). More importantly, the ways in which humans communicate have been found to be based on the ways in which animals communicate, so that learning about the latter has proven useful when it comes to learning about the former (Sterelny, 2016; Jackendoff, 1999; but see also Mithen, 2005). If this is true for language, there is at least some reason to think it will also be true for economic decision-making.

This is further supported by the third reason for resisting the view that economic decision-making is a uniquely human trait: the fact that (economic) decision-making appears to be a trait with a major adaptive importance for *many* organisms—including, but not restricted to, humans. For example, there is no doubt that many animals have to decide whether to consume food resources now or later, whether to share them with others, and how to best allocate their time among different possible uses (collecting food, eating food, mating, etc.) (Hagen et al., 2012; A.I. Houston & McNamara, 1999; Kalenscher & van Wingerden, 2011; Santos & Rosati, 2015; Kenrick et al., 2009). Moreover, there is no question that these are some of the most adaptively important behaviors these animals need to engage in (A.I. Houston & McNamara, 1999; Alcock, 2013). Hence, many animals plausibly also have evolved the ability to make economic decisions.

Of course, this is not to say that there are no unique features of or adaptive pressures on human economic decision-making (see the following and Chapter 5 for more on this). In general, the details of the adaptive problems faced by different groups of organisms will differ slightly from each other. The point here is just that it is implausible that economic decision-making in humans is so unique that it completely invalidates the use of comparative methods.[33]

Overall, therefore, it becomes clear that it is at least prima facie reasonable to appeal to data on animal decision-making to better understand human economic decision-making. However, exactly how does this appeal work? In particular, keeping in mind the discussion of the previous section, exactly how can we use data about animal economic decision-making to make progress in answering questions [1] and [2] [a]/[b]? To make this clearer, the rest of this section develops a methodological framework with which to interpret the actual data on animal economic decision-making so as to make it possible to relate it to these two questions. Developing such a framework is especially useful in cases where the actual data are only recently acquired, hard to interpret, and therefore subject to revision—which, as will also be made clearer in the discussion that follows, is exactly the situation in the present context.

To develop this framework, it is useful to begin by considering the remarks of Kalenscher and van Wingerden (2011, p. 8), who provide the following sketch of how data on animal economic decision-making can be useful for our understanding of human economic decision-making.

> We regard the critical examination of economic theory the prime objective of performing experiments in decision making and we maintain that in this light it is imperative to include animal models in the arsenal at our disposal. In the worst case, results obtained in animals will corroborate those obtained from humans, strengthening the existing theory. Preferably, though, comparative research will uncover inconsistencies in choice behavior between humans and animals that allow for an improved, more comprehensive description of choice behavior and possibly force us to re-think the basis of economic theory in the light of the evolutionary roots of choice.
>
> Kalenscher and van Wingerden (2011, p. 8)

While containing important insights, it turns out that Kalenscher and van Wingerden's (2011) methodological thoughts need to be refined and corrected in several different ways.

To see this, it is best to follow Kalenscher and van Wingerden (2011) and distinguish several possibilities of what the data on animal economic decision-making could show. However, unlike them, it turns out to be most useful to make a *three*-part division of what these data could be like: the data on animal economic decision-making could exactly and uniformly mirror the ones from human economic decision-making, there could be overall mirroring with some variation in animal economic decision-making, and there could be some uniqueness in human economic decision-making. Consider these three cases in turn.

1. Perfect Mirroring

Assume it turned out that all animals make economic decisions (to the extent they do so at all—i.e. to the extend they are representational decision makers: see Schulz, 2018b) exactly like humans. What would this show?

To answer this question, note first that Kalenscher and van Wingerden's (2011) suggestion cannot be quite right as put. This case cannot "corroborate [the results] obtained from humans," thus "strengthening the existing theory" simply because there is no *one* existing theory here. In fact, as noted previously, this is precisely the problem. There are several different theories competing for acceptance, either as the sole theory describing human economic decision-making or as part of a pluralistic theory.[34] This, though, does not mean that Kalenscher and van Wingerden (2011) are wrong in thinking that finding that animals make decisions exactly like humans do is helpful for our understanding of the nature of (human) economic decision-making.

The reason for this is that this scenario of perfect mirroring in animal and human economic decision-making provides evidence (nothing more, but also nothing less) for the *pluralist* position concerning (human) economic decision-making. That is, if it turned out that (the relevant set of) animals, too, make some decisions that can be predicted well using RCT, some decisions that can be predicted well using PT, some decisions that can be predicted well using RT, and some decisions that can be predicted well using SH, this suggests that economic decision-making in humans (and animals) is pluralistic in nature. This can be seen from noting that, given the—as argued previously, reasonable—presumption that economic decision-making in animals is comparable to that in humans, perfect mirroring suggests that all of the different theories of economic decision-making get at *something* about the shared psychological foundations surrounding human and animal economic decision-making. After all, if there were *one* right theory describing all economic decision-making, we would expect that its superiority would show up *sometime*. If, even after expanding the data set to include data on animal economic decision-making, it continues to be the case that all theories are about equally good descriptions of economic decision-making—i.e. if their predictive accuracy remains equally high (Hitchcock & Sober, 2004)—this suggests that they should all be taken about equally seriously.

Note that this last point would not hold if it were not the case that all of these theories make different predictions that have been confirmed. It is not just that all the theories can accommodate all the data equally well; it is that they make substantively different predictions, all of which continue to be confirmed (Hitchcock & Sober, 2004). There is just no reason to think that these different predictions will continue to be confirmed as more data are obtained—unless the different theories get at different aspects of reality. For these reasons, the case of perfect mirroring suggests that the answer to question [1] is: *human—and animal—economic decision-making mechanism is subserved by a disunified collection of different economic decision-making mechanisms.*

However, when it comes to question [2][b]—i.e. the domains of application of the different economic decision-making mechanisms making up this collection—things are not so straightforward. Perfect mirroring between human and animal economic decision-making will, at best, *constrain* the answers we can give to this question, but it does not itself make a positive suggestion as to how to answer it. The reason for this is easy to see. Given that, in the human case, we are uncertain as to how to carve up and categorize the domains of the different economic decision-making mechanisms, much of that uncertainty will just carry over to the exactly mirrored version of these data in the animal case. What we may be able to rule out are hypotheses that locate the domain of application of a given type of decision-making mechanism (the RCT-like one, say) in some *distinctively human* sphere. For example, if we find that animals, too, make economic decisions in an RCT-like way (at least sometimes), the domain of application of the latter has to be shared between them and humans, and thus cannot be something uniquely human (such as one concerning language-based economic interactions) (see also Santos & Rosati, 2015). However, beyond ruling out hypotheses that locate the domain of application of a particular decision-making mechanism in some distinctively human sphere, complete mirroring in animal and human economic decision-making does not help us to answer question [2][b]. This scenario just does not provide the "inconsistencies in choice behavior" (Kalenscher & van Wingerden, 2011, p. 8) that provide information with which to answer this question. For the latter, we need to design further experiments (on humans or animals) that can address this question specifically.

All in all, the case of perfect mirroring provides evidence for the pluralistic answer to question [1], but merely constrains answers to question [2]. Consider next what happens in the case of imperfect mirroring—when there is variation in the ways that animals make economic decisions.

2. Mirroring With Variation in Animal Decision-making

Assume it turns out that the decisions of some animals—chimps, say—can be well predicted just by relying on RCT; those of others—gorillas, say—just by relying on PT; those of a third type—orangutans, say—just by relying on RT; and those of a fourth type—macaques, say—just by relying on SH. Overall, there seems to be mirroring in the way in which humans and animals make economic decisions, but this mirroring is imperfect. While no single type of animal makes economic decisions like humans do, taken together, animals do make decisions exactly like humans do. What are we to say about this case?

In regard to question [1], this case yields the same answer as the one of perfect mirroring: it provides evidence for the pluralistic picture of human economic decision-making. As previously, if human economic decision-making were really best described by *one* of RCT, PT, RT, or SH, then we would expect this to become clear as we obtain more data. However, the fact that we continue to find support for all of these different theories even when we broaden the data and also consider animal economic decision-making suggests that these different theories really get at different aspects of a pluralistic decision-making mechanism. Indeed, the fact that *different* animals make economic decisions in ways that favor *some* of these theories over others—i.e. that, as in the previous example, the decisions of chimps can be well predicted just using RCT, and those of macaques just using SH—makes this evidence even stronger than in the case of perfect mirroring. After all, this variation in animal economic decision-making suggests that the different ways of making economic decisions are *dissociable*. This further underwrites the hypothesis that RCT, PT, RT, and SH are descriptions of *genuinely different* decision-making mechanisms—which just happen to all be present in humans (Santos & Rosati, 2015, pp. 9, 11–13, 18).

More than that, though, the variation in the ways in which animals make economic decisions provides crucial extra information that can be used to provide evidence for how to answer question [2][b] (Santos & Rosati, 2015). The fact that different animals can be differentially well predicted to make decisions using RCT, PT, RT, and SH makes it easier to triangulate on the environments that trigger the relevant decision-making mechanisms. For example, if it turns out that SH is a good predictor of the decisions of macaques and RCT a good predictor of the decisions of chimpanzees, this suggests that the domain of application of SH is an environment shared between humans and macaques and RCT an environment shared between humans and chimpanzees.

It is important to emphasize again that this is not guaranteed. It is possible that different groups of organisms make economic decisions in the same way for very different reasons, and thus that the domains of application of a given way of making economic decisions differ among these organisms (Santos & Rosati, 2015; Kalenscher & van Wingerden, 2011). However, this does not invalidate the fact that, at least *ceteris paribus*, the existence of shared ways of making economic decisions in different organisms speaks in favor of these shared ways being adaptations to their common environments (Santos & Rosati, 2015; Kalenscher & van

Wingerden, 2011). In turn, this suggests that it is these common environments that mark the domain of application for this shared way of making decisions—thus taking a step towards answering question [2][b]. (I return to this point in the following sections.)

In all then, mirroring of human and animal economic decision-making, in combination with the existence of variation in animal economic decision-making, provides more information than in the case of perfect mirroring. In particular, this sort of case can strengthen the pluralistic answer to question [1] and provide some means for answering question [2][b]. Consider then the final case: the existence of unique features of human economic decision-making.

3. The Existence of Unique Features of Human Economic Decision-making

The last case to be considered here is the scenario in which there is only partial overlap in the ways humans and animals make economic decisions: while both humans and animals show decision-making features F_1-F_n, only humans show F_{n+1}-F_{n+m}.[35] (To make the discussion clearer, I assume that there is no variability in the ways animals make economic decisions—the only difference is between the ways in which humans and animals make economic decisions. However, all the conclusions carry over where there is such variability, though the issues of the previous sub-section then need to be addressed as well.) So, for example, perhaps it turns out that animals only act in ways that suggest expected utility maximization and only humans display the endowment effect, appear to make intransitive choices, or make decisions in ways that suggest reliance on simple heuristics like "Take the First" or "Take the Best." What are we to conclude about questions [1] and [2][a]/[b] here?

In this extreme case, the inference should be towards a monistic answer to question [1]: the extra data from animals show that, overall, RCT is the best description of human and animal economic decision-making. To see this, recall that it is taken for granted that all theories can, somehow, make sense of all of the *human* data. If we add to this the (supposed) facts that (a) humans and animals make economic decisions in comparable ways, and (b) animals are clearly shown to be relying on an RCT-like economic decision-making mechanism only, this then gives us reason to think that humans do so, too. Here, the extra data on animal economic decision-making resolve the ambiguities in the human data, and show that an RCT-like mechanism is at the heart of human and animal economic decision-making (Kalenscher & van Wingerden, 2011, p. 6).[36] In this way, we obtain evidence for, on the one hand, a monistic answer to question [1], and, on the other, an RCT-focused answer to question [2][a].

Effectively, the same points hold for the more general case in which animal economic decision-making is describable in several different ways as well, but where these ways are fewer in number than in the human case. In this more general case, we do not get evidence for a full-on *monistic* answer to question [1]. Rather, in line with the previous remarks concerning both mirroring and human uniqueness, we get evidence in favor of a *limited form of pluralism*—one that leaves out the feature that is uniquely human. The reasoning leading up to this conclusion is exactly the same as before: the animal data provide a way to cut through the tangle of human data.[37] As for how to answer question [2][b], this case does not differ from the previous ones: it depends on whether there is variability among animal economic decision-making.

So, for a concrete example, assume animals can be well predicted (in different circumstances, perhaps) to make decisions by just relying on RCT, PT, and RT, but humans are unique in *also* making some decisions in ways that can be especially well predicted using SH. Here, the appropriate conclusion to draw is that these kinds of data are evidence for the fact that economic decision-making is constituted by a pluralistic collection of RCT-like, PT-like, and RT-like—but *not* SH-like—decision-making mechanisms. We might then further go on to determine the domains of application of the RCT-like, PT-like, and RT-like decision-making mechanisms by seeing which animals share these mechanisms with humans, and in which environments these live.

The natural next question to answer is why there is ambiguity in the human data that is absent in the animal data in the first place. Why can human economic decision-making be well described in an RCT-like, PT-like, RT-like, *or* SH-like manner, when this is not true for animal economic decision-making, which is describable with only a subset of these theories? What accounts for the difference here?

In response, note that, for a variety of reasons, human economic decision-making is likely to be enmeshed with many other issues (Kalenscher & van Wingerden, 2011, p. 6; Hutchinson & Gigerenzer, 2005, p. 118). For example, due to the fact that humans communicate linguistically, they face a whole slew of decision situations that animals do not face—whether to accept or make a marriage proposal, take out a mortgage or make a loan, or retweet the latest internet meme. Each of these cases plausibly adds a number of complexities apart from the pure economic decision problem at stake: what being married entails in the particular culture in question, whether being indebted comes with possible social stigmatization in the culture in question, and what is considered funny in the culture in question (Bshary & Raihani, 2017; Kalenscher & van Wingerden, 2011). In turn, the fact that human economic decision-making is thus embedded in many other issues plausibly makes it harder to determine the economic "signal" out of the surrounding "noise." By contrast, animal economic decision-making can be seen as a "cleaner" version of human economic decision-making, whose nature can thus be more easily ascertained (see also Kagel et al., 1995).

Two further complexities surrounding the present scenario of some human uniqueness in economic decision-making need to be mentioned. First, nothing in this analysis is meant to rule out the possibility that there have been unique evolutionary pressures in the human case—whether selective or not—that led to a novel set of economic decision-making mechanisms in that lineage. The important point to note here is just that the existence of these unique evolutionary pressures raises separate evolutionary biological considerations from the consideration of data concerning animal economic decision-making. If we can identify the evolutionary pressures on human economic decision-making directly, then there is little need to appeal to data on animal economic decision-making. However, if we cannot do so, the consideration of animal economic decision-making continues to be evidentially relevant for the answer to questions [1] and [2][a]. (At any rate, as will be made clearer in the next section, it looks as though the discussion of this situation is moot anyway.)

The second complexity to be discussed here concerns the case in which the human uniqueness does not concern the presence or absence of one of these (putative) decision-making mechanisms, but merely its extent. So, perhaps, animals, like humans, can be seen to make economic decisions in an RCT-like, PT-like, RT-like, and SH-like manner—it is just that humans appear to make relatively *more* decisions in a PT-like manner than animals. That is, what if the uniqueness of human economic decision-making does not concern the nature of its aspects, but merely in the relative centrality of these different aspects?[38] In that case, this situation is best seen as a special case of imperfect mirroring: it speaks in favor of a pluralist answer to question [1]. What makes this case special is that it can provide further evidence as to how to answer question [2][b]: the human-animal differences in the centrality of the different components of the economic decision-making machinery can suggest what the domains of application are of these different components (Santos & Rosati, 2015). So, if we find that humans make PT-like decisions relatively more often than chimps, we may be able to hone in on the environments that differ between these two cases, and thus identify the domain of application of PT-like decision-making. This point will become important—and be further illustrated—momentarily.

All in all, this thus yields the following. Kalenscher and van Wingerden (2011) are right that data about animal economic decision-making can be evidentially useful for determining the most compelling theory of human economic decision-making. However, the issues here are significantly more complex than they let on. If the animal data exactly mirror the human data, we do not get confirmation of the existing theory (as there is no such theory). Rather, we get some support for a pluralistic picture of human and animal economic decision-making—but, except for in the case in which there is some human uniqueness in the centrality of the different aspects of this pluralistic mechanism, relatively little guidance as to how the different parts of the pluralism work together (i.e. what their domains of application are). In the case of imperfect mirroring, we still get support for a pluralistic picture of human and animal economic decision-making, but also more guidance as to how the different parts of the pluralism work together. Finally, in the case where there are unique features in human economic decision-making, we get support for a reduced form of pluralism—or even monism. This all can be summarized with the following table:

Table 4.2 A framework for bridging non-human and human economic decision-making

	Consequences for Q[1]	*Consequences for Q[2] [a]/[b]*
Perfect mirroring	Pluralism (weaker evidence)	Domains of application cannot be uniquely human.
Mirroring with variation/ Human uniqueness (in extent only)	Pluralism (stronger evidence)	Domains of application are the shared environments of organisms that can be well predicted using the relevant theory (RCT/PT/RT/SH).
Human uniqueness (in kind)	Monism/Limited pluralism	Domains of application are the shared environments of the relevant organisms (strength of inference depends on whether there is variation in animal economic decision-making).

With this methodological framework in hand, it is now possible to consider what the actual data about animal economic decision making imply for how to answer questions [1] and [2][a]/[b].

IV. Interpreting the Actual Data on Animal Economic Decision-making

When looking at the actual data on animal decision-making, the first and major point to note is that, as things stand, there is overwhelming evidence for some form of *mirroring* between human and animal economic decision-making (Santos & Rosati, 2015; see also Hutchinson & Gigerenzer, 2005). While there is much that we do not know about the ways in which different animals make economic decisions, what we know so far suggests that they make them much like humans do. The following table summarizes some of the relevant findings here:[39]

Table 4.3 Findings about animal economic decision-making

Decision-making Feature	Animal	Source
EU maximization	Macaques, rats, pigeons	Glimcher et al. (2005); Kagel et al. (1995); Santos and Chen (2009)
Framing effects, risk aversion in gains/risk love in losses; endowment effect	Chimpanzees, gorillas, orangutans, Capuchin monkeys, rats, honeybees, starlings	Chen et al. (2006); Lakshminarayanan et al. (2008); Brosnan et al. (2007); Brosnan et al. (2008); Drayton et al. (2013); Flemming et al. (2012); Kacelnik and Bateson (1996); MacDonald et al. (1991); Marsh and Kacelnik (2002); Shafir et al. (2008); Shafir (1994)
Intransitive decision-making/ counterfactual reasoning	Macaques, chimpanzees, bonobos, grey jays	Abe and Lee (2011); Lee et al. (2005); Hayden et al. (2009); Rosati and Hare (2013); Waite (2001)
Reliance on simple heuristics	Wasps, bumblebees, flies, grouse	Goulson (2000); Karsai and Penzes (2000); Shettleworth (2009); Bonduriansky (2003); Gibson (1996)

By this methodological framework, therefore, actual data on animal economic decision-making suggest that the answer to question [1] is a pluralist one. This, by itself, is a key point to emphasize: taking into account what we know about animal economic decision-making provides good evidence for the claim that economic decision-making—in humans and animals—is not constituted by a unitary, monistic mechanism, but by a pluralistic collection of different mechanisms. Put differently, the reason why there is ongoing debate between defenders of RCT, PT, RT, and SH is that they are *all* right—decision-making is best seen as constituted by a number of different aspects (see also Kenrick et al., 2008; Fiske, 1992). (Interestingly, this conclusion receives independent support from the consideration of the evolutionary pressures on representational decision-making: Schulz, 2018b, chap. 8.)

However, it is not yet fully clear whether all animals make economic decisions like humans do, or whether there is some variability in the ways in which animals make economic decisions. Indeed, in some cases, the data on animal economic decision-making are somewhat difficult to interpret. For example, while there is evidence of frustration when chimps and bonobos do not get the most valuable decision outcome, whether that is full-fledged regret—and thus a case of RT-based decision-making—is not yet entirely clear (Rosati & Hare, 2013, pp. 6–7; Zeelenberg, 1999; Santos & Rosati, 2015). For this reason, it is not yet fully clear how question [2][b] is to be answered. The actual data on animal economic decision-making do not tell us a whole lot about how the different aspects of the human/animal economic decision-making machinery are coordinated— i.e. which of these aspects is being triggered in which circumstances.[40]

That said, these data do give some *clues* about how to answer question [2][b]. There are two reasons for this. In the first place, there is some human uniqueness that these data do establish. This uniqueness, though, is of the minor variety that concerns differences in the centrality of different aspects of this pluralistic decision-making machinery. The major point that has been established here is that the endowment effect, while generally shared with other animals, seems uniquely widespread in the human case (Kanngiesser et al., 2011; Brosnan et al., 2012). This is important to note, as there is a good explanation for this divergence. As Santos and Rosati (2015) note, while animals can make a distinction between what is theirs and what is someone else's, this distinction is rooted in quite concrete terms. By contrast, humans can think about resources more abstractly, and thus can extend their dispositions to value property more widely than other animals. This thus suggests that the domain of application of PT—for which the endowment effect is central—is decisions concerning what to do with *property*: whether to acquire it, or whether to part with it. By contrast, strategic, interpersonal decisions, investment and savings decisions, intertemporal discounting, and occupational decisions are likely to be taken in a different way. This is also a relatively precise prediction that can be tested further.

The second hint about how to answer question [2][b] concerns the fact that SH-like decision-making has been found in birds and insects (see e.g. Goulson, 2000; Shettleworth, 2009; Gibson, 1996). This matters, as humans, birds, and insects live in quite different social environments. In particular, while keeping track of the social roles and psychological states of other organisms is widely seen to be crucial to human living (and characteristic of great ape environments in general—see e.g. Humphrey, 1986; Whiten & Byrne, 1997; Sterelny, 2003, 2012a; Schulz, 2018b), this is not generally thought to be crucial for birds or insects. While the latter can also be social, their sociality is quite different (S. R. Griffin et al., 2012; Seeley et al., 2006; Karsai & Penzes, 2000). In turn, this suggests that social decision-making—where this is understood as concerning the kinds of social environments the great apes inhabit—may not be the domain of simple heuristics. This is also something that confirms the view of other researchers (Sterelny, 2003; Schulz, 2018b; but see also Hurley, 2005; Gigerenzer, 2007).

In this way, we can use the data on animal economic decision-making not just to underwrite a pluralist answer to question [1], but also to suggest, in response to question [2][b], that (i) the domain of application of the PT-like decision-making mechanism concerns property-focused decisions, and (ii) the domain of application of the SH-like decision-making mechanism excludes social (in the great ape sense) decision-making. In this way, the data on animal economic decision-making can be seen to make a limited, but still very useful contribution to the debates surrounding human economic decision-making. Two final points are important to note about this conclusion.

First, it needs to be acknowledged that future studies of animal economic decision-making could find further divergences in human and animal economic decision-making—or show that the divergences we thought existed (such as concerning the endowment effect) do not, in fact, exist. For this reason, this conclusion should not be thought to have a finality that it in fact lacks: this conclusion needs to be seen to be relativized to the current state of knowledge of animal economic decision-making. However, this does not make this conclusion uninteresting—especially given the evidential focus of the present form of evolutionary economics. Indeed, this is precisely why this chapter developed a *framework* for relating data on animal economic decision-making to debates about human economic decision-making. While it looks as though animal economic decision-making is a more or less perfect mirror of human economic decision-making, it may well turn out that, with further data, human economic decision-making is actually unique and thus monistically structured in key ways.

Second, it also needs to be noted that the data on animal decision-making referred to here draw on many different types of animals. This leaves it open that a more restricted set of comparison animals would yield different results from the ones mentioned here. Indeed, as noted earlier, the inference from how organism O_1 makes decisions to how organism O_2 makes decisions is generally taken to be stronger the more closely related the two organisms are. For this sort of reason, Brosnan et al. (2011) (for example) have suggested that the old-world monkeys might make for more relevant comparison organisms than new world monkeys for human economic decision-making.

However, this suggestion does not in fact change the gist of the conclusions described previously. First, many of the major patterns in human decision-making are found in even a restricted set of animals—e.g. just the old-world monkeys, or even just the great apes (Santos & Rosati, 2015). Second, there is no clear cut-off as to how closely related the organisms need to be to make meaningful comparison classes (Felsenstein, 2004; Harvey & Pagel, 1991). So, it is known that some traits are inherited from deep within the phylogenetic tree (as is true, e.g., of some aspects of human neural organization—Grillner et al., 2013; Greene, 2008). Third, this kind of suggestion merely reinforces the point just made that these conclusions are relativized to our current understanding of animal economic decision-making. So, it is of course entirely conceivable that, on the one hand, we discover reasons for excluding some types of animals as relevant comparison classes for human economic decision-making, and, on the other, we thereby find divergences that favor a more restrictedly pluralistic—or even monistic—picture of human economic decision-making. This is thus another reason why developing this framework is important and useful.

V. Conclusions

I have considered the question of whether and when data from animal decision-making can be useful for addressing key economic debates. Specifically, I have focused on the debate about the most compelling economic theory of choice (especially among RCT, PT, RT, and SH), and came to three overarching conclusions. First, what the economic debate among these different economic theories of choice comes down to is the questions of whether economic decision-making is constituted by a monistic or a pluralistic decision-making mechanism, and of what the nature of this mechanism is. Second, I have developed a methodological framework for relating data on animal economic decision-making to these questions. According to this framework, mirroring among human and animal economic decision-making speaks in favor of a pluralistic view of economic decision-making, and imperfect mirroring especially can suggest the domains of application of the different components of this pluralistic economic decision-making machinery. By contrast, the discovery of significant human uniqueness speaks in favor of a restricted form of pluralism, or even monism. Third, and finally, I have argued that the available data about animal decision-making suggest that humans and animals make decisions in very similar ways, thus favoring a pluralistic view of economic decision-making. Further, these data suggest that PT-like decision-making—which is especially pronounced in humans—is focused on property decisions, and that SH-like decision-making—which humans share with birds (among others)—is not focused on social (in a great ape sense) decisions.

Therefore, I hope to have shown that data about animal thinking and acting *can* be evidentially useful for making progress in economics—even if only in a limited manner. However, I also hope to have shown that these data need to be interpreted appropriately: relating them to the disputes about human economic decision-making is far from straightforward.

Suggested Further Reading

For some classic defenses of why economic decision-making should be placed in an evolutionary biological framework, see Gigerenzer and Selten (2001), V. Smith (2007), and Stanovich (2004). A related view applied to decision-making especially in financial contexts is Lo (2017). Important works investigating economic decision-making in non-human animals are Santos and Chen (2009), Santos and Rosati (2015), Kalenscher and van Wingerden (2011), DeAngelo and Brosnan (2013), and Brosnan et al. (2012). Boyer (2008) presents evolutionary biological reasons for the human tendency to hyperbolically discount the future.

Notes

1. Other social and natural sciences are also concerned with the ways in which people make decisions, but economics is unique in its strong focus on this issue—indeed, some authors have *defined* economics as the science of decision making (see e.g. Mill, 1874; Robbins, 1932; Kalenscher & van Wingerden, 2011, p. 1). Note also that it is contentious whether and how economic decisions can be distinguished from other decisions (see e.g. Becker, 1976; Hausman, 2012). However, as long as it is acknowledged that non-human animals make decisions that can be classified as economic decisions—which it should be, as also made clearer in the following—nothing here depends on how this question is answered. For this reason, in what follows, I use the terms "economic decision making" and "decision making" interchangeably.

2. The term "comparative methods" can be understood narrowly or broadly. In the narrow sense, it refers to methods that use the comparison among different groups of organisms to understand the selective pressures on traits of (some of) these organisms (Santos & Rosati, 2015). In the broad sense, it refers to methods that use the comparison among different groups of organisms to understand a more general set of questions about the traits of (some of) these organisms—including, but not limited to, their selective history (Felsenstein, 2004; Harvey & Pagel, 1991; Schulz, 2013). This is sense in which this term is used here.

3. Note that the focus here is strictly on the question of what data on animal economic decision making can tell us about human economic decision making. There is no doubt that these data can add much of value to various other questions—most obviously, they deepen our understanding of animal cognition (which we care about for its own sake). However, the latter is not central here.

4. A quick word about terminology. There are many different labels attached to the kinds of theories of choice offered by economists: "rational choice theory," "decision theory," "preference theory," "utility theory," etc. However, as will be made clearer momentarily, many of these labels can be somewhat misleading—e.g. economic theories of choice need not be seen as concerning matters of rationality and they need not involve utilities. For this reason, I opt here for the more neutral terminology of "economic theories of choice."

5. As also noted in Chapter 1, references to "mechanism" here should not be taken to imply commitment to the specific way of understanding this term argued for e.g. by Bechtel and Abrahamsen (2005) or Machamer et al. (2000), but more generally as referring to the psychological/neurobiological structures underlying (economic) decision making.

6. This is sometimes expressed by saying that the notion of "preference" involved in economics is that of "revealed preferences" only. However, as pointed out by Hausman (2012), this can be a somewhat misleading label. At any rate, the point is that on this reading of the aims of the economic study of choice is just the prediction of choice behaviors, not their psychological underpinnings or rational defensibility.

7. This distinction has come to be somewhat controversial—see e.g. Shapiro (2004) and McClamrock (1991). However, it is appealed to here purely for expository purposes (the point in the text can be reformulated in other theoretical frameworks), and so no further discussion of it is necessary in the present context.

8. It is possible to subdivide the normative reading of the aims of the economic theories of choice into a behavioral and a psychological/neurophysiological variant: on the former, it is the aim of these theories to specify the kinds of behaviors agents should engage in so as to be rational, whereas on the

latter, it is the aim of these theories to specify the kinds of mental processes agents should undergo so as to be rational. However, as will be made clearer momentarily, making this finer distinction is not necessary here.

9. Of course, as teleosemantic theories have shown, facts about what has evolved can ground *some* normative claims. The point here is just that it is not obvious how they can ground claims *about the proper theory of rationality*. For more on teleosemantics, see e.g. Millikan (1989, 1984) and Papineau (1987).

10. Some may prefer to see the detailed specification of the psychological structures underlying choice behavior as a matter of psychology or cognitive neuroscience, and not economics (see e.g. Gul & Pesendorfer, 2008). However, as noted earlier in Chapter 1, this is neither plausibly motivated, nor amounts to much more than a linguistic point: there is an interesting question to be addressed here, and whether the question is seen as "economic" or "psychological" in nature is really secondary.

11. This glosses over the debate between causal and evidential forms of RCT, for which these formalisms would be slightly different (see e.g. Eells, 1982; Joyce, 1999; Jeffrey, 1983). Note also that instead of seeing u(x) to be a minimally cardinal function, we can also talk about a *set* U of different utility functions. Finally, note that as suggested in Chapter 2 it is a fundamental result of RCT that the ranking based on expected utility values is equivalent to a specific kind of ordering over the actions the agent believes to be open to her—e.g. an ordering that is complete, transitive, and perhaps satisfying some kind of independence axiom (the details of which kinds of assumptions the ordering satisfies differ from version to version of RCT). However, this point is not so central for present purposes. Similarly, there is debate over how the utility function discounts future states of affairs compared to present ones; however, again, this is not so central here. For more on this, see e.g. Gowdy et al. (2013) and Boyer (2008).

12. This also glosses over some differences in different forms of RCT, where the nature of these probabilities differ—i.e. whether they are subjective or objective. See also Resnik (1987). However, this difference is not so important here, as on the positive interpretations of RCT, an appeal to objective probability needs to be understood as underwritten by a "Principal Principle" (Schaffer, 2003; Lewis, 1980) of sorts: agents' subjective probabilities match the objective probabilities (and where there are no objective probabilities, a different theory of choice needs to be appealed to).

13. PT is sometimes seen as merely describing *the choice patterns that result from* the reliance on heuristics and biases in decision making (see e.g. Okasha, 2011). However, there is also no question that PT can be and has been interpreted as itself describing the mechanisms underlying economic decision making (see e.g. Santos & Rosati, 2015). This is how it will be understood here.

14. A specific function proposed by Tversky and Kahneman (1992) is $\pi(x) = p(x)^\delta / (p(x)^\delta + (1 - p(x))^\delta)^{1/\delta}$, where p(x) is the probability of state of world x occurring. However, the details of this function are not so important here.

15. Note that rank-dependent utility theory (see e.g. Quiggin, 1993; Schmeidler, 1989; Epstein, 1999; Abdellaoui, 2002) can be seen as a version of this kind of theory as well, as it, too, is based on seeing agents as evaluating not just the consequences of a given action, but how these consequences relate to the other possible actions (Diecidue & Wakker, 2001). This is noteworthy also, as rank-dependent utility theory is often associated with PT, given that it has been incorporated into a version of the latter (Tversky & Kahneman, 1992). However, in the present context, it is better seen as another way of spelling out the core idea behind RT (Diecidue & Wakker, 2001).

16. Q(x) can be expanded beyond pairwise choices (Loomes & Sugden, 1987; Bleichrodt & Wakker, 2015), but for present purposes, the restricted version in the text is all that is needed. Finally, a more general version of RT abstracts away from utility differences and lets Q(x) range over different options more generally. Again, though, the restricted version in the text is all that is needed for present purposes.

17. As pointed out previously (note 13), the work surrounding PT is sometimes interpreted in this vein as well. See also V. Smith (2007).

18. Note that this sense of simplicity is organism-relative: which decision rules it is easy to use will depend on the cognitive abilities of the agent in question. Note also that not all heuristics need to be simple in this sense. For example, MCMC analysis is a heuristic for determining probability distributions, but its computational and temporal requirements are high.

19. This relates to the work of Simon (1957).

20. Of course, one could argue that plausible standards of rationality must have some empirical relevance (Lyons et al., 1992; Boudry et al., 2015; Polonioli, 2015). However, this would then entail seeing RCT at least partly as a positive theory.

21. RCT is also the key economic theory of choice expounded on in most economic textbooks (see e.g. Mas-Colell et al., 1996; Varian, 1992). However, this may be mostly due to pedagogical and/or historical-sociological reasons, so I do not consider it in detail here.

22. It may be thought that it is possible to make progress here by considering factors other than those that fit to the data, such as parsimony (Hitchcock & Sober, 2004; Sober, 2015). However, given that the competitors here are not clearly comparable along dimensions of parameter number (say) and given that—as I make clearer later—the different theories here need not be seen as competitors, this is not greatly promising. For attempts to do something like this in a related area, see Schulz (2018a). See also note 32.

23. There are many other examples here that could be cited as well, but for present purposes, focusing on these four is sufficient—especially in light of the data on animal decision making cited here, which focus on these four cases as well. See also V. L. Smith (2005, 2008) and Guala (2005) for more on experimental economics.

24. The distinction between prediction and accommodation will become important again in the following and in the next chapter.

25. Risk aversion in gains/risk love in losses is "built into" PT. However, since the theory then predicts a number of other related findings (such as endowment effect), further findings of risk aversion in gains/risk love in losses can be taken as evidence for PT.

26. In some versions of RCT—such as that of Savage (1954)—this can lead to some awkward cases, as every consequence is assumed to be able to be combined with every action. This means that it needs to be possible—in principle, though not necessarily in practice—to bring about the consequence "being $100 richer after being given $200 and accepting a 50–50 gamble of losing $100 or losing $200" by doing the action "not accepting the gamble." This does not affect all versions of RCT, though—e.g. Jeffrey (1983) is not affected by it—and the extent to which this is really a problem can be debated as well.

27. Furthermore, to the extent that this is not an available strategy here, then defenders of PT can also appeal to changes in the way the agent evaluates the prospects open to her; this is parallel to what is true for RCT in the following.

28. As noted in Chapter 2, this may be a case of adaptive preferences (but need not).

29. Note that there is often interpersonal variation in these domains: not all humans make all decisions in the same ways (Binmore, 2007; Stanovich & West, 2000). The next chapter looks at this issue in more detail. For now, though, the point to note is just that the different theories of choice are differentially successful in predicting data across different decision domains.
30. Alternatively, the different theories may be seen as different modeling approaches towards an underlying unitary reality. However, as long as it is granted that it is an open question whether all of these different modeling approaches are useful—and if so, when—this comes out to the same as the claim made in the text in the present context (See Phattanasri et al., 2006, for a similar idea in a different context.). (Thanks to Colin Allen for useful discussion of this point.)
31. This could thus be seen as a meta-version of SH. It is not just that different decision situations lead to the use of different simple heuristics, but some decision situations lead to the use of decision rules that are not simple at all— namely, ones that are RCT-like, PT-like, or RT-like (Hutchinson & Gigerenzer, 2005, p. 101). See also Schulz (2018b, chap. 8) for more on this.
32. Note also that it is not clear if the pluralistic theory is more complex than any of RCT, PT, RT, or SH—and if so, exactly how much more complex it is. This is due to the fact that the pluralistic theory cannot be a simple "addition" of RCT, PT, RT, and SH—these need to be *integrated* appropriately (Machery, 2009). Hence, it is not straightforward to use model selection theory here (Hitchcock & Sober, 2004) to make progress in assessing these issues. At any rate, the point in the text is that independently of questions of theory complexity other sources of evidence are available that if properly interpreted can help make progress resolving these issues.
33. Here, it is also important to keep in mind that as made clear in Chapter 1 nothing more than the *evidential* relevance of evolutionary biological considerations in economics is argued for. So, it need not be the case that the study of animal decision making answers all open questions concerning human decision making—i.e. that there is nothing unique about human economic decision making. As long as animal decision making can be shown to have *some* evidential relevance to human economic decision making, that is all that is needed here.
34. The reason why Kalenscher and van Wingerden (2011) assume that there is one theory is that they take canonical RCT to be the normative baseline, from which various departures have been observed in the human case. However, as noted earlier, this is not a useful way of conceiving the dispute here; this conflates normative and descriptive approaches towards the study of economic choice and fails to do justice to the many existing alternative theories of economic choice.
35. Note that depending on how features are coded this could mean that humans make decisions in a more or less complex way compared to animals. I return to this point later.
36. Glimcher et al. (2005) present a view a bit like this.
37. Indeed, the extreme case of there being only one way of making decisions that is shared between humans and animals can be seen as the limiting version of the present case, where the "pluralism" that is being supported is restricted to one decision making mechanism.
38. If the ways in which humans and animals make decisions is very finely individuated, this case is bound to occur. It is overwhelmingly plausible that there are *some* differences between human and animal economic decision making. The issue here, though, is whether these differences concern systematic

divergences in the extent to which animal decision making is well described as relying on RCT, PT, RT, or SH.

39. Note that this table does not represent a thorough compilation of all the relevant findings here; it is just meant to illustrate the gist of these findings. Note also that there are a number of findings in which there is mirroring between humans and animals that do not distinguish among the different theories of contention here (such as concerning hyperbolic discounting or ambiguity aversion). These are also not further considered here. See Santos and Rosati (2015) and Kalenscher and van Wingerden (2011) for more on these.

40. Similarly, the strength of the answer to question [1] is not fully clear either.

5 Not All the Same

The Selection-Based Approach Towards Economic Decision-making (The Evidential Project II)

I. Introduction

There is a growing body of evidence that human economic decision-making is not a unitary trait (even if, as argued in the previous chapter, it is pluralistic in nature). There appear to be systematic differences in the ways in which different groups of humans make economic decisions (see e.g. Byrnes et al., 1999; Henrich et al., 2005, 2001).[1] What we do not yet know is whether these documented differences are *fundamental* or merely *superficial*. That is, what is still unclear is which (if any) of the systematic differences in the ways in which different humans make economic decisions are just the product of differences in the decision situations faced by different humans, and which (if any) are differences in the ways different humans are psychologically structured. The aim in this chapter is to address this question in more detail: is the psychological structure of economic decision-making a human universal (though one that leads to surface-level differences if employed in different decision situations), or is this structure variable?

It is clear that answering this question matters from an anthropological perspective. For its own sake, we care about determining how different groups of humans think and act—and economic decision-making is part of this project (Winterhaler, 2007; Henrich et al., 2005, 2001). However, answering these questions is also important from an economic point of view (see also Downes, 2016a, 2016b). If different groups of humans make economic decisions fundamentally differently, then the investigation of economic decision-making needs to be careful in its choice of study subjects, as well as in the scope of the conclusions it draws from these studies (for a clear statement of this point, see e.g. Henrich, 2000, p. 973).

It is also clear that addressing this issue is, to a large extent, a matter of empirical studies concerning the ways in which different groups of humans make decisions. In particular, the more persistent the differences in economic decision-making that have been found turn out to be—even if every effort is made to control for differences in the decision situations in the different humans in question—the more reasonable it is to see these differences as fundamental (Henrich, 2000; Henrich et al., 2005, 2001). However, as will be further made clear in the following, this does not mean that an appeal to considerations from evolutionary biology might not be useful here as well. On the one hand, there may be much uncertainty remaining about the nature of the empirical findings—how many of them are artifacts of the study design, how many are statistical noise, and how are they to be interpreted in the first place? On the other hand, even given the possibility to investigate the situation empirically, the appeal to evolutionary biology promises to make available a different and novel source of evidence concerning this issue (Hammerstein & Stevens, 2012). Since more data is (generally) better than less data, this thus provides further support for an appeal to evolutionary biology.

In this context, it is also noteworthy that this issue makes for a telling example of quite a different way in which evolutionary biological considerations can be evidentially brought to bear on the resolution of economic questions. In the previous chapter, the focus was on comparative methods and the use of animal models in the investigation of human economic decision-making. In the present context, by contrast, the focus is on the identification of specific evolutionary pressures on *human* economic decision-making (see also Collins et al., 2016). That is, the goal here is to determine whether there are clues directly in the evolutionary history of humans that should lead us to expect that there is—or is not— diversity in the fundamental mechanisms underlying human economic decision-making. As will be made clearer throughout this chapter, this focus on the evolutionary history of humans brings out a number of novel aspects of the evidential evolutionary economic project more generally.[2]

The chapter is structured as follows. In section II, I make the question to be addressed clearer: what diversity in the ways in which humans make economic decisions has been found, and how, in principle, can this diversity be accounted for? In section III, I lay out what is methodologically required in order to underwrite compelling evidential arguments for or against the fundamentality of human diversity in economic decision-making. In section IV, I apply the standards laid out in the previous section to some arguments for or against the fundamentality of different specific forms of diversity in human economic decision-making. I conclude in section V.

II. Diversity in Human Economic Decision-making: Evidence and Explanatory Schemes

It is becoming increasingly clear that there are differences in the ways in which different groups of humans make economic decisions. The goal of this section is, on the one hand, to illustrate the kind of diversity in human economic decision-making that has been found, and on the other, to lay out some of the ways in which this diversity can be evolutionary biologically explained.

1. Two Examples of Diversity in Human Economic Decision-making

In what follows, I set out two examples of human diversity in economic decision-making. These examples cover (a) gender differences and (b) cultural differences in economic decision-making. The choice of these examples is driven by three considerations: breadth, centrality in the literature, and expository usefulness.

In particular, the aim is first to give a sense of the breadth of diversity in economic decision-making that has been found. For this reason, the issues covered concern both *abstract features* of economic decision-making— the attitudes towards risk inherent in it—as well as the *content* of the principles people employ in their economic decision-making—the extent to which they are disposed to share resources. While falling short of the full gamut of differences that have been and are being explored (which also include, for example, gender and/or cultural differences in the time horizon of economic decision-making, the extent to which people value international trade and its products, and the disposition towards innovation: see e.g. Y. J. Li et al., 2012; Boyer & Petersen, 2017; Boyer, 2008; Y. J. Li et al., in press; Baumard, in press), this is sufficient to at least give a *sense* of this gamut.

Second, the aim is to pay particular attention to issues that have received especially strong attention in the literature: here, gender and cultural differences on the one hand, and differences in attitudes towards risk and sharing on the other loom large. For this reason, the present focus on gender differences in attitudes towards risk and cultural differences in attitudes towards sharing, while limited, does at least get at a substantial part of the core of the existing discussion on this topic.

Third, the aim in what follows is to consider examples that make it easy to bring out the promises and challenges in evidentially connecting evolutionary biology and economics. As will become clearer in the sections to follow, the discussion of these two issues is very useful in this regard.

With this in mind, consider these two examples in turn. More details about them will be provided momentarily; for now, the goal is just to make the ensuing discussion clearer and more concrete.

a. Gender Differences in Attitudes Towards Risk

Economists assess an agent's attitudes towards risk by considering how she evaluates gambles with the same expected value but different degrees of variance. As also noted in Chapter 2, if the agent evaluates an uncertain option by just paying attention to its expected value—without considering its variance—she is said to be risk-neutral; if she is drawn towards options with lower variance, she is said to be risk-averse; and if she is drawn towards options with greater variance, she is said to be risk-loving (Mas-Colell et al., 1996). Now, it turns out that there is evidence that, while both males and females show both risk aversion and risk love (shown differently for gains and losses, as noted in the previous chapter), males often seem to be relatively more risk-loving compared to females (Byrnes et al., 1999; Eckel & Grossman, 2008).

So, field studies have shown that males place more "high-risk" bets than females (J. E. V. Johnson & Powell, 1994) and that the retirement savings portfolios of females are more risk-averse than those of males (Bajtelsmit & VanDerhei, 1997; Jianakoplos & Bernasek, 1998; Sundén & Surette, 1998). Some laboratory studies further confirm these data—it appears that males tend to be less risk-averse than females (Eckel & Grossman, 2008; Powell & Ansic, 1997).[3] It needs to be stressed from the get-go, though, that these gender differences are generally quite labile, weak, and need to be interpreted with a lot of care (J. A. Nelson, 2015); this is an important point to which I return later. Still, for now, the key point to note is that there is evidence for the contention that males tend to be (slightly) more risk-loving than females (across a variety of measures of attitudes towards risk).

Of crucial importance for what follows further is the fact that these differences in attitudes towards risk do not seem to be restricted to specific decision situations, but concern choices about monetary gambles, retirement portfolios, and many other decision situations. That is, what needs to be explained here is why females and males frequently seem to display different attitudes towards risk, not just when it comes to a particular choice situation, but in many different types of choice situations.

b. Cultural Differences in the Propensity to Share

In economics, it is not per se assumed that people are egoistic in resources: what goes into people's utility functions, prospect functions, regret functions, or what cues their simple heuristics are attuned to is typically left open (at least a priori) (Hausman, 2012). (That said, it is at least a working implicit assumption of much work in economics that people prefer more of a good to less, and that they are not greatly disposed towards reducing their own consumption to increase that of others: Mas-Colell et al., 1996; Hausman, 1992.) What is surprising, however, is that the empirical literature on the extent to which people are disposed to share resources with others—especially in settings involving the ultimatum game or various public goods games—has uncovered massive cross-cultural differences (Henrich et al., 2005).

For example, it has been found that, in some cultures, minimal offers in the ultimatum game are frequently made and accepted, whereas in other cultures even hyper-fair offers (i.e. offers of more than half of the available resources) are rejected (Henrich, 2000; Henrich et al., 2005, 2010, 2001; Bone et al., 2016; Blake et al., 2015). More generally, there seem to be major cultural differences in how people evaluate the world when it comes to decisions involving both self and others. Different cultures seem to be associated with very different utility (etc.) functions (on average). In some, people typically share a lot, and in some less so. This is unsuspected from an economic point of view and requires an explanation (Henrich et al., 2005).

2. Two Ways of Explaining Diversity in Economic Decision-making

Differences in economic decision-making (like the ones just sketched) could be *fundamental* or *evoked*.[4] Fundamental differences are due to differences in the in the psychological makeup of different humans: people make different economic decisions simply because they are different sorts of economic decision makers. Evoked differences, by contrast, are due to environmental differences that, in combination with a shared underlying psychology, lead agents to make systematically different economic decisions.[5] Figure 5.1 makes this clearer:

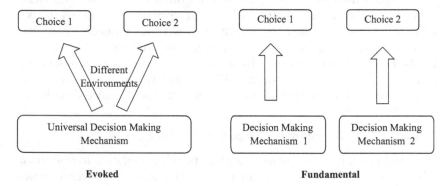

Figure 5.1 Evoked vs. fundamental explanations of human diversity

As has been noted earlier, the overriding interest in this distinction in the present context lies in the fact that it has major implications for the nature of economic theorizing. If differences in human economic decision-making are merely *evoked*, then economic theories about decision-making (whether pluralistic or not) will apply across the board—at least as far as the differences in question are concerned. This would put these theories on par with theories concerning a number of other biological processes, such as the biological mechanisms underlying liver functioning (Trefts et al., 2017). In turn, this implies that the relevant differences in economic decision-making need not be taken to require relativization of economic theories. The more similar the decision situations of different agents can be made to be (in the relevant regards), the more similar the decisions will be that these agents end up making.

If, by contrast, the stated differences in economic decision-making (or others like them) turn out to be fundamental, then successful economic theorizing *does* need to be relativized to different groups of humans. The relevant differences in economic decision-making will persist even if the agents are put in the same decision situations. Different genders and cultures would need to be seen as different kinds of economic decision makers (at least in certain key regards). As Henrich (2000, p. 978) puts it, in cases like this, "the implicit assumption that all humans share the same economic decision-making processes, the same sense of fairness, and/or the same taste for punishment must be brought into question." This would put economic theories of choice on par with theories concerning a number of other sociocultural phenomena, such as politeness. There is no one theory of polite behavior; rather, any theory of politeness needs to be relativized to the group of humans in question.

It is the goal of the rest of this sub-section to make these two ways of explaining differences in economic decision-making clearer. Before doing this, though, three points are worth emphasizing.

First, there is no need to assume that all differences in economic decision-making must be given the same kind of explanation. It is entirely possible that some differences in economic decision-making are evoked and others fundamental.

Second, drawing the distinction between fundamental and evoked explanations of human differences in economic decision-making depends on being able to determine when two decision-making mechanisms are tokens of the same type, and when they are of different types. That is, drawing this distinction depends on a plausible individuation schema for economic decision-making mechanisms. This is an important point to which I return when laying out the evoked account in more detail.

Third, evolutionary biological considerations are at the heart of the distinction between fundamental and evoked explanations of differences in economic decision-making. Whether a difference in human economic decision-making is fundamental or evoked depends on whether humans evolved in such a way that there is interspecific variation in economic decision-making, or whether they evolved in such a way that economic decision-making is, on a "deep" level at least, a human universal. Hence, looking at evolutionary biological considerations makes for an at least prima facie plausible inroad into the assessment of whether a given difference in economic decision-making is fundamental or evoked.

a. Fundamental Explanations of Diversity in Human Economic Decision-making

Fundamental diversity in human economic decision-making—where it exists—stems from the fact that, for one reason or another, humans have evolved in such a way that there are systematic differences in the biological mechanisms underlying economic decision-making. What factors, though, could lead to this kind of divergent evolution? Several answers to this question have been proposed; the following is an outline of a number of them. While clearly there is much more that could be said about each of these answers, a relatively brief sketch is sufficient for present purposes.

(1) DIFFERENT SELECTION PRESSURES

Most straightforwardly, it is possible that different populations of human beings have been subject to different selection pressures for significant periods of time, leading to the evolution of different traits.[6] For example, it has been argued that the differences in the production of hemoglobin in Tibetan highlanders as compared to other humans can be explained by the fact that Tibetan highlanders have lived at high altitudes for significant periods of time, and have thus evolved specific adaptations for dealing with this (Beall et al., 2010; Huerta-Sanchez et al., 2014). Also, David Buss and his colleagues (see e.g. D. M. Buss & Schmitt, 1993) have argued that, for much of their evolutionary history, males and females have faced different selection pressures when it comes to their sexual strategies, which explains the (alleged) observable differences in these strategies (a point that will become important again in the following).[7]

(2) FREQUENCY-DEPENDENT SELECTION, SELECTION FOR DIVERSITY, AND DRIFT

The second biological explanation of human diversity appeals to the fact that evolutionary pressures can be more complex than involving (mostly or only) directional selection, and directly favor—or at least lead to—diversity. There are many different variants of this explanation; for simplicity, I focus on three of them (see also Nettle, 2007).

In the first instance, it is possible that humans evolved in environments that led to frequency-dependent selection of a given trait. This happens when the fitness of the trait is a function of how common that trait is in the population. A major example is the hawk/dove game (Maynard Smith, 1982). In this scenario, what should be expected to evolve is a combination of hawks and doves: neither hawk nor dove is an evolutionary stable strategy—the evolutionarily stable state will contain both hawks and doves.[8] What this means is that humans could differ in a significant dimension (such as how aggressive they are) not because they faced different selection pressures, but because they faced the same, frequency-dependent selection pressures.[9]

The second version of this kind of evolutionary biological explanation of human diversity concerns cases when there is selection for diversity itself. A famous case of this is the idea that the diversity in some of the genes responsible for regulating immune responses in humans may be due to the fact that this diversity itself was adaptive: e.g. because it made it harder for viruses or other pathogens to attack human populations (see e.g. Prugnolle et al., 2005; Hughes & Yeager, 1998).[10]

The third scenario to be considered here concerns cases of drift. In particular, it may be that some traits are selectively largely neutral, so that, by drift, different alleles of these traits went to fixation in different human populations. A purely illustrative example of this may be differences in aesthetic preferences (though see also the discussion in Chapter 2). It may be that whether a person prefers fine lines and cold tones to thicker lines and warmer tones—or the other way around—has no effect on her expected reproductive success. Given this, it may happen that, purely by chance, one of these preferences goes into fixation in one human population, and another in another (Gillespie, 1998). In other words, some human differences may not be due to selection pressures of one kind or another, but to the *absence* of any kind of selection pressure.[11]

(3) GENE-CULTURE COEVOLUTION AND THE EVOLUTION OF IMITATION AND LEARNING

It is by now widely accepted that humans have evolved minds that are prone to imitate or learn from others (see e.g. Heyes, 2018; Boyd & Richerson, 2005). The reason for the evolution of this disposition to imitate and learn is thought to lie in the fact that it may make it easier for different human populations to adapt to the locally prevailing conditions (Boyd & Richerson, 2005; Henrich, 2015). So, for example, it may be that different agricultural practices are differentially adaptive in different environments. If humans have a disposition to imitate and learn from others, then it can, under some conditions, be easier and quicker for them to discover the locally optimal agricultural practices. This is important, as it implies that this sort of scenario can explain cases of human diversity that may be out of reach of the slower, non-cultural evolutionary biological explanations (Boyd & Richerson, 2005).

Importantly, moreover, the gene-culture coevolutionary approach can also explain cases of human diversity that are not well explained by appealing to what is adaptive at all (e.g. because they feature different locally *maladaptive* variants). The reason for this is that once a general disposition to imitate or learn from others has evolved, it can support human differences of many different kinds (Boyd & Richerson, 2005). For example, perhaps people are particularly drawn to imitate successful

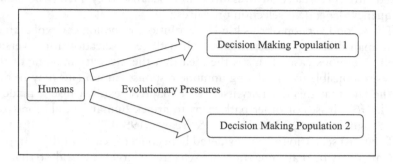

Figure 5.2 The evolution of fundamental diversity in economic decision-making

people, or their parents, or some combination of these. If there is some stochasticity in who is successful and in what different parents do, different human populations can end up looking very differently—simply because different practices happen to have been imitated (for detailed models of this sort, see e.g. Boyd & Richerson, 2005).[12]

In all, then, fundamental explanations of diversity in human economic decision-making see these differences as resulting from the fact that humans faced (possibly frequency-dependent) selective, non-selective, or gene-cultural pressures towards making economic decisions in different ways. Figure 5.2 makes this clearer.

b. Evoked Explanations of Diversity in Human Economic Decision-making

An interspecific difference is said to be evoked if it is due to the fact that the same underlying biological machinery leads to different traits in different environments (see e.g. Tooby & Cosmides, 1992; Hittinger & Carroll, 2007; Penke, 2010). So, for example, genetic clones of the corn plant can grow to different heights depending on whether they are in a nutrient-rich or nutrient-poor soil. This difference in heights is evoked, as underlying it are not differences in the biological mechanisms of plant growth, but differences in the *inputs* to these mechanisms only.[13]

From an evolutionary biological perspective, such evoked explanations of individual differences are often taken to be a sort of "null model." Unless a reason to think otherwise is provided (such as one of the fundamental explanations just cited), it is assumed that, *fundamentally*, organisms of the same taxon have the same set of traits, with any differences in *displayed* traits being due the interaction of this same set of fundamental traits with the organisms' different environments (Tooby & Cosmides, 1992). In this context, this implies that it is presumed that all humans are fundamentally the same; only if specific empirical or theoretical reasons to think otherwise are provided is this assumption dropped. While there is room to question how plausible this kind of null-model assumption is (D. M. Buss & Hawley, 2010; Buller, 2005), for present purposes, it does not need to be discussed further (though I return to this issue momentarily).

At any rate, it is at least sometimes possible to provide positive reasons for why a given difference in human economic decision-making (or any other trait) should be seen to be evoked. In the main, these reasons center on the fact that it may sometimes be possible to underwrite the idea that all of the relevant subpopulations—males and females, say, or different cultures—were subject to the same evolutionary pressures. So, perhaps there is reason to think that all of these subpopulations were subject to directional (gene-cultural) selection of the same kind. Or perhaps they were all subject to the same kind of stabilizing (gene-cultural) selection. Or perhaps all of these sub-populations faced sufficient amounts of inter-migration that any differences between them were being "washed out" (Sober & Wilson, 1998).

Thus, by relying either on the null-model assumption or on positive arguments, it can be possible to provide reasons—nothing more, but also nothing less—to think that humans evolved so as to rely on the same decision-making mechanisms in the relevant regards. Figure 5.3 makes this clearer.

Two further (related) points concerning this characterization of evoked differences need to be made. First, it is necessary to place some restrictions on the form that evoked differences can take, as this category otherwise threatens to become so broad as to include *all* cases of diversity. Genetic differences, for example, could be seen as just different environmental triggers that lead the same biological machinery—meiosis, mitosis, and gene expression—to produce different traits. The same goes for differences that are the result of learning: different learning environments in combination with the same learning machinery can lead to different outcomes. This, though, would fail to be theoretically interesting, as it would no longer provide meaningful subdivisions among cases of interspecific diversity. Even if there are theoretical contexts in which all cases of interspecific differences merely differ in degrees, that does not mean that it is not useful to divide them into separate categories for some *other* theoretical purposes. In particular, whether a difference counts as evoked

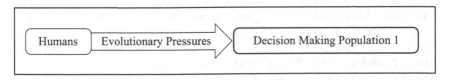

Figure 5.3 The evolution of merely evoked diversity in economic decision-making

or fundamental depends on how the relevant biological mechanisms are to be individuated. In the case of the corn mentioned earlier, if it is predictively or explanatorily reasonable to see corn height as determined by a biological mechanism that takes nutrients as an input, this difference counts as evoked; by contrast, if it is predictively or explanatorily reasonable to see corn height as determined by a biological mechanism that is (partly) constituted by the soil the plant is in, then this difference counts as fundamental. These different answers can be differentially plausible in different investigative contexts.

Importantly—and this is the second point to note here—the notion of fundamentality/"evokedness" that is at stake in the present context needs to dovetail with the best way of theorizing about *economic decision-making*. Recall that the question to be addressed here is whether economists need to be open to the fact that their theories need to be relativized to different subsets of the human species, or whether one set of theories (whether pluralistic or not) can accommodate all of human economic decision-making. Hence, in this context, the best way of individuating biological mechanisms—and thus, of distinguishing evoked from fundamental differences—is in relation to the economic decision-making mechanisms at stake in the different cases. If it turns out that different humans rely on different decision-making apparatuses, then that is a reason to see the resultant differences in their choice behaviors as fundamental in nature—independently of the question of whether the differences in these mechanisms should themselves be considered fundamental or evoked.[14] Evoked explanations of (systematic) differences in economic decision-making, *as these are understood here*, therefore see these differences as stemming from the same economic decision-making machinery reacting to (systematic) differences in the decision situations encountered by different human populations.

III. Methodological Constraints in the Evolutionary Biological Investigation of Diversity in Human Economic Decision-making

Before considering some of the specific evolutionary biological arguments that have been given for one or another type of explanation of human diversity in economic decision-making—fundamental or evoked— it is necessary consider an immediate objection to the usefulness and/ or possibility of taking an evolutionary biological approach towards this issue. This objection is based on the idea that providing compelling and complete accounts of the evolutionary pressures on human economic decision-making may be too difficult to be useful (Richardson, 2007; Brandon, 1990; Buller, 2005). It may be possible to say something about the evolution of economic decision-making *in general* (as shown in the previous chapter). However—or so it may be thought—precisely determining the *specific* evolutionary pressures on economic decision-making *in the human lineage* is epistemically (nearly) impossible (Richardson, 2007; Brandon, 1990; Buller, 2005). Two broad sets of reasons for this skeptical view are typically given.

First, setting out a complete human-specific evolutionary biological approach towards the explanation of differences in economic decision-making would seem to require a thorough understanding of the genetics underlying human economic decision-making. This will tell us, on the one hand, if and how this trait was heritable, and on the other, if there were pleiotropic effects or epigenetic interactions that complicated its spread.[15]

Second, setting out a complete human-specific evolutionary biological approach towards the explanation of differences in economic decision-making would seem to require an understanding of the population structure and size of the human populations economic decision-making evolved in. This is needed, as it will tell us about the relative existence and strengths of drift or various group-level influences on the evolution of this trait (see also Sober, 2008; Sober & Wilson, 1998).

Obtaining this kind of knowledge, though, seems very difficult. After all, human economic decision-making is a trait that does not fossilize and which is likely to have a complex genetic basis. On top of this, many aspects of human evolutionary history are still unknown, including many aspects of the structure and size of ancient human populations. For these reasons, one may be skeptical that providing an account of the evolution of specifically human economic decision-making is possible (Richardson, 2007; Brandon, 1990).

However, drawing this kind of skeptical conclusion would be too quick. Despite the difficulties just mentioned, the appeal to evolutionary biological considerations in the discussion of the explanation of human diversity in economic decision-making should not be discounted as impossible (see also Saad, in press; Downes, 2013).[16] In the main, this is because a *full* account of the evolution of human economic decision-making is (or at least need) not be the goal here (though it would of course be nice to have). As also noted in Chapter 1, what we are looking for are just well-supported *partial* accounts. The goal is to obtain *evidence* for the evoked or fundamental explanations of some case of diversity in human economic decision-making, not full *confirmation* of any particular explanation. For the former, evidentialist picture, though, many of these issues can be left open.

So, while it is true that the genetic underpinnings of a given trait can bias its evolution positively or negatively, this does not mean that determining the selective pressures on that trait (for example) in the absence of full knowledge of these genetic underpinnings is completely uninteresting. In particular, the determination of the selective pressures on the trait provides *one reason* for why it evolved in a certain way. If this reason turns out to be weakened (or strengthened) by a different, genetics-based reason, this does not mean that establishing this first, selection-based reason is entirely uninteresting. Much the same goes for the other issues mentioned earlier, such as the presence of significant migration among different human populations, or the fact that these populations were very small. These factors certainly can impact the evolution of a trait. However, this does not mean that determining the other evolutionary pressures on that trait is uninteresting in the absence of knowledge of these factors.

Indeed, as noted in Chapter 1, this is a familiar point about scientific inference. Much of science is about building up a picture of the nature of some phenomenon using a large number of small pieces of evidence. Each of these pieces may be "drowned out" as the inquiry progresses, but until this happens these pieces are our best estimate for what the overall picture looks like (Sober, 2008; Schulz, 2018b). Because of this, the appeal to evolutionary biological considerations in the investigation of human diversity in economic decision-making should not be dismissed out of hand as a non-starter: it can just be seen as providing *evidence* for the sources of this diversity.

However, this does not mean that there are no constraints on a partial, evidential evolutionary biological treatment of human diversity in economic decision-making. In particular, despite the fact that such an appeal need not be able to provide a full account of the evolutionary pressures on human economic decision-making, it cannot be *purely* speculative either. Even partial evolutionary accounts are subject to two important constraints (see also Wylie, 2002, 1994; Schulz, 2018b).[17]

First, any account of the evolutionary pressures on human economic decision-making must be *internally* coherent. That is, it must at least be possible to spell out the account in such a way that it *could* be seen as a reason to favor evoked or fundamental explanations of some divergence in economic decision-making. This may seem trivial, but can in fact be hard to satisfy. For significantly complex accounts with many different elements, it is not always obvious how all of the elements work together.

Second, any account of the evolutionary pressures on human economic decision-making must be *externally* coherent. That is, it must at least be consistent with other known sources of evidence or modes of inference relevant to this trait. The fact that the goal need not to be to provide a full evolutionary treatment of the trait in question should not be taken to imply that known findings relevant to the evolution of the trait can be *ignored*. There is a difference between a partial account and an empirically unsupported account.

Note that formally spelling out these notions of "coherence" is not trivial (see e.g. Thagard, 2000; Bovens & Hartmann, 2004, for some attempts). Fortunately, for present purposes, a detailed treatment is not necessary. All that matters here is the informal sketch: proposed evolutionary explanations of diversity in human economic decision-making need to be able to actually speak in favor of an evoked or fundamental account, and they should match what else is known about human evolution and its investigation. Nothing more precise than this is needed for what follows (see also Wylie, 2002, 1994).

IV. Two Evolutionary Biological Arguments Concerning Human Diversity in Economic Decision-making

To illustrate the promises and challenges of appealing to evolutionary biological considerations in investigating human diversity in economic decision-making, it is useful to now consider two specific arguments—one concerning each of the two instances of human diversity in economic decision-making sketched in section II.

1. Gender Differences in Attitudes Towards Risk

How is the fact (assuming that it is indeed a fact) that human males show greater risk love compared to females to be explained? To answer this question, it first needs to be made clearer what the difference between the fundamental and the evoked accounts amounts to in this context.

Given the decision-making pluralism argued for in the last chapter, there are many possible sources of a fundamental difference in attitudes towards risk among females and males. It could be the case that males and females differ in terms of the circumstances in which they rely on RCT-like, PT-like, RT-like, or SH-like economic decision-making mechanisms—i.e. *when* they rely on one of these mechanisms rather than another one. However, it could also be that the *nature* of the RCT-like, PT-like, RT-like, or SH-like economic decision-making mechanisms are different in males and females. For example, they could rely on different utility functions, different prospect-valuation functions, different p-functions, different Q-functions, or different simple heuristics. Finally, both could be true at the same time.

In contrast, the evoked account of differences in attitudes towards risk among females and males holds that males and females tend to systemically face slightly different decision situations, so that, while they rely on the same kinds of RCT-like, PT-like, RT-like, or SH-like economic decision-making mechanisms—i.e. mechanisms based on the same utility functions, prospect-valuation functions, p-functions, Q-functions, or simple heuristics—they differ in the extent to which they are drawn towards high-risk choices. In other words, the displayed differences in attitudes towards risk among females and males are not seen as stemming from differences in the actual mechanisms underlying economic choices in males and females, but as merely due to systematic differences in the decision situations faced by them.

To make the discussion in the rest of this chapter tractable, the focus will be on the question of whether females and males rely on different utility functions, prospect-valuation functions, p functions, Q functions, or simple heuristics, or whether they rely on the same such functions and heuristics, but face different decision situations such that, in combination with the nature of these (shared) functions and heuristics, their displayed attitudes towards risk differ. That is, I shall not consider the question of whether females and males (also) differ in terms of the situations in which they rely on the different aspects of their pluralistic decision-making machinery. This more restricted focus has the benefit of being in line with the evolutionary biological arguments that have in fact been given in this context, which tend to focus on exactly this contrast. At any rate, the present, limited perspective is sufficient for the methodological purposes that are at the heart of this chapter (and the book as a whole).

With all of this in the background, consider the major evolutionary biological argument for an *evoked* account of gender differences in attitudes towards risk that has been put forward in the literature.[18] This argument follows the "null model" assumption of evoked explanations of human diversity and presupposes that the utility, prospect-valuation functions, p-functions, etc. underlying economic decision-making do not differ among the genders. What the argument tries to establish is that there are evolutionary biological reasons for thinking that the decision situations faced by females and males are systematically different in key dimensions, so that the attitudes towards risk *displayed by* the two genders should be expected to be different—despite the fact that their underlying decision-making mechanisms are not.

To do this, this argument relies on the fact that "minimal parental investment"—the resources a person minimally needs to invest in order to reproduce—is higher for females than for males (Y. J. Li et al., 2012; D. M. Buss & Schmitt, 1993; Trivers, 1972; Daly & Wilson, 1983; N. P. Li & Kenrick, 2006; Kenrick et al., 1990).[19] Reproduction for males only requires the time to achieve a mating, plus the costs of making the sperm cells. By contrast, females not only need to spend time in achieving the mating and making the egg cell, but they also face a 40-week long pregnancy with major physiological changes, many missed mating opportunities, and a relatively dangerous birthing process. In turn, this means that, for any given mating opportunity, the costs for females are higher if the mating is of a "bad" type—the mate has bad genes, does not provide help in raising the offspring, or comes with few other adaptive benefits—than for males (for which these costs are often practically zero).

To understand this better, it is necessary to consider a widely accepted distinction in the time horizon of human mating (see e.g. D. M. Buss & Schmitt, 1993; N. P. Li & Kenrick, 2006). In long-term mating, the goal is to find a partner for a long-term relationship. This partner can help in raising children as well as make the pursuit of social and physical resources easier and more efficient (D. M. Buss & Schmitt, 1993). By contrast, in short-term mating, the goal is to find a partner for just one or a few matings in a brief period of time (N. P. Li & Kenrick, 2006).[20]

While, in many organisms, both males and females engage in both kinds of mating strategies (D. M. Buss & Schmitt, 1993; Cockburn, 2013), differences in minimal parental investment are particularly important in the short-term. In the long-term, it can also be in the male's adaptive interest to invest considerably in the offspring, as such offspring can be significantly fitter than non-invested offspring—which thus increases the male's own fitness (Sober, 2001). This is quite plausible in the human case, where offspring are very dependent on their caregivers for a long period of time, and where cultural learning and social intelligence—both of which can be provided by males as well as females—are drastically important to their success in surviving and reproducing (Henrich, 2015; Sterelny, 2012a; D. M. Buss & Schmitt, 1993; Heyes, 2018).

In the short-term, though, differences in minimal parental investment do matter. For males, short-term mating opportunities mostly come with just upsides: they may obtain extra offspring at virtually no cost to themselves (D. M. Buss & Schmitt, 1993). There may be no guarantee that the female will in fact raise the offspring, but even a small chance of this happening can outweigh the often near zero costs of engaging in the mating.[21] However, for females, the situation is different. Given the significant amount of minimal parental investment they need to provide, short-term matings come with major potential costs. While they do also get the adaptive benefit of having the offspring (should they choose to raise it), they get that benefit minus the significant costs of the minimal parental investment they have to expend.

In turn, this makes it unclear why it would it be in the females' interest to engage in short-term mating, rather than just looking for long-term partners. In response, several different answers have been proposed (D. M. Buss & Schmitt, 1993; Gangestad & Simpson, 2000). The key one among these is that short-term matings provide females with the chance to obtain offspring with particularly good genes (Gangestad & Simpson, 2000; D. M. Buss & Schmitt, 1993; Pillsworth & Haselton, 2006). If it is possible for females to identify which males are likely to be especially well-endowed genetically, it can be in their adaptive interest to engage in short-term matings with these males *even if* these males are unlikely to help in raising the offspring. (This will be so especially if it is possible for them to defray the costs of raising the offspring—e.g. by having a long-term partner invest in the offspring as if it were their own.) After all, the adaptive value of having offspring is an increasing function of the genetic endowment of the offspring: healthier, stronger, more attractive offspring are likely to yield more grand offspring (Sober, 2001; Gangestad & Simpson, 2000). Furthermore, it is not implausible that females can identify, at least to some extent, which males are likely to have especially good genes: for these good genes are likely to have external effects in the male—making him healthier, stronger, and more attractive—that can be detected (though possibly with a margin of error) by females (Gangestad & Simpson, 2000; D. M. Buss & Schmitt, 1993; Pillsworth & Haselton, 2006).

It is important to note that this "good genes"-based account of the reasons for why females should engage in short matings may need to be amended in certain ways (see e.g. Mays & Hill, 2004). However, for present purposes, this is not so important. What matters here is just that, on the one hand, females may well have some evolutionary, "good-gene" related reasons—whatever, exactly, these turn out to be—to engage in short-term mating, but that on the other hand, these reasons are significantly balanced by the downsides of short-term mating. In turn, this implies that, for females, a randomly selected male is less likely to make for an adaptive short-term partner that for males. For males, virtually all females make good short-term partners, whereas for females, only a small subset of males do.

This matters, as it can be used to provide an evolutionary biological argument for the fact that females and males should be expected to sometimes *display* different attitudes towards risk, despite (the null-model assumption of) there being no fundamental difference in their decision-making machinery. This argument begins by assuming that the economic decision-making machinery relied on by females—i.e. the utility function, prospect-valuation function, p-function, Q function, or simple heuristics they use to make decisions—is correlated with their fitness. Note that this assumption is one aspect of the natural selection/economic decision-making parallelism discussed in Chapter 2: humans (whether female or male) should be expected to make decisions in ways that are at least correlated with what is biologically advantageous for them. Now, as noted in Chapter 2, this assumption is substantive and cannot always be expected to hold; however, as also noted in Chapter 2, the assumption can be expected to hold at least sometimes. For present purposes, it is best to start by accepting it—I return to its plausibility later. Given this assumption and the minimal parental investment difference noted previously, there is reason to think that females and males will typically face mating options with very different *net* fitness values, and thus, very different likelihoods of being chosen. This matters, as these differences in the net values of the same types of mating options can be used to underwrite differences in attitudes towards risk. To make this clearer, consider figure 5.4.

Figure 5.4 represents the situation where a given female or male has to decide whether to engage in a short mating with a potential mate whose gross mate value is uncertain, and can be either "high" or "low." Here, the gross mate value is the value of the mate without the costs of the mating—i.e. without taking into account the minimal parental investment the female or male has to pay when engaging in short-term matings. (For

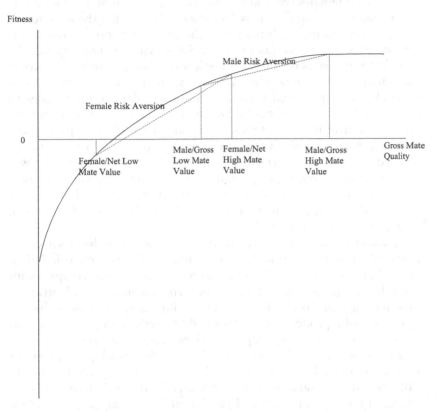

Figure 5.4 Fitness function for males and females for short-term mates with
different gross mate values

example, the gross mate value may be the quality of the mate's genes.) By
contrast, the net mate value is the gross mate value minus the costs of the
mating in terms of minimal parental investment. Since minimal parental
investment is close to zero for males, this means that the gross mate value
is very close to the net mate value for males, whereas for females, who do
face substantial amounts of minimal parental investment, the net mate
value for a potential mate will generally be lower than the gross mate
value—and can be negative if the males has a low gross mate value to
begin with.

   What is important about this case is that it implies that the decision situation of females and males is systematically different, even for
partners with the same gross mate value (e.g. with genes of the same
quality). Females will tend to face decisions among significantly worse
options than males: the average short-term mate for a female will have
an expected net mate value that will be much lower—and may in fact be
negative—than the average short-term mate for a male. This matters, as

it implies that females will likely turn down more gambles across short-term mates than males, as it is less often adaptive for them to accept such gambles. Assuming further that the fitness function across short-term matings is concave (as in Figure 5.4)—which is not implausible, given that improvements in the "genetic value" of a mate likely become quite small at the margin—this also suggests that, at least when it comes to the choice of potential mates, it would be adaptive for females to be more risk-averse than males. For males, "gambles" over short-term mates are nearly linear in fitness—they are on a "flatter" part of the fitness function over short-term matings. By contrast, for females, the expected value of an uncertain mating opportunity is significantly lower than the value of a mating opportunity with the same expected outcome, just guaranteed—they are on a more "curved" part of the fitness function over short-term matings (Y. J. Li et al., 2012; Gneezy & Leonard, 2009, p. 1654, footnote 18).[22]

Importantly, these facts may further be thought to have knock-on effects on the behavior of males. Given that females are "choosier" than males when it comes to short-term matings, males have to compete with each other to get access to females. In turn, this makes it adaptive for them to engage in daring, dangerous behavior. First, given the biological importance of reproduction, males should be willing to pay a very high price to obtain the chance to reproduce. Not every male gets to reproduce, so that risking much to be one of the ones that does is adaptive. (This is a classic point in the literature on sexual selection—see e.g. Clutton-Brock, 2007.) Second, engaging in some types of risky, daring behaviors can be a signal of the male's mental and physical abilities. Only strong, healthy, males can engage in (potentially even frivolous) behaviors that come with a high chance of injury (Y. J. Li et al., 2012; D. M. Buss & Schmitt, 1993; Zahavi, 1977; Miller, 2001; M. Wilson & Daly, 1985).

Hence, overall, the fact that females pay a minimal parental investment premium over males should lead us to expect that they will (at least sometimes) act as if they are more risk-averse than males. Importantly, this is so despite the fact that, underlying their decisions are the same fundamental attitudes towards risk.[23]

However, a closer look at this argument shows that it fails to be sufficiently internally coherent to make for genuine evidence in the *present* debate surrounding the explanation of gender differences in attitudes towards risk. First, it needs to be noted that, as already hinted at, this argument rests on some questionable assumptions: the existence of a link between the fitness of a decision option and its utility/prospects/regrets/simple heuristics cannot be taken for granted (as shown in Chapter 2), and the "null model" assumption of no differences across the genders unless reasons to the contrary are produced is also not without its detractors (as noted earlier). The fact that this argument relies on these—not obviously true—assumptions is seen to weaken it from the start. However,

I will bracket further discussion of this point here, and concentrate on the second set of issues to be raised with the present argument.

This second set of issues concerns the fact that this argument fails to connect in the appropriate way to the conclusion it is meant to derive. Recall that the issue to be explained is why females and males seem to display different attitudes towards risk *in many different contexts*—including non-mating ones. The data seem to suggest that there are differences in the displayed attitudes towards risk by the two genders in a number of different circumstances, including (especially) financial ones. There is no question that these data are very complex, hard to interpret, and not always pushing in the same direction (a point that has been noted earlier and which will also be made clearer momentarily). However, it is also clear that these data do not just concern mating decisions. It is not the case that we find gender differences in displayed attitudes towards risk in mating decisions, and not in other contexts.[24] Rather, we find gender differences in displayed attitudes towards risk across the board— though there is not always consistency as to exactly which contexts these differences can be found in and how stable these differences are. What this means is that any compelling evolutionary biological explanation of gender differences in displayed attitudes towards risk—whether evoked or fundamental—needs to be able to speak to the existence of such differences *outside of mating contexts, too*.

However, this is exactly what the previous argument *cannot* do: it can only underwrite differences in displayed attitudes towards risk *about mating decisions*. This is so for two reasons. First, this argument is based on the existence of a gender asymmetry in minimal parental investment. However, this kind of asymmetry is only *directly* relevant for mating decisions: it concerns features of the results of the decision to reproduce. Second, the only route towards explaining a wider set of gender differences in displayed attitudes towards risk is by appeal to the male intrasexual competition resulting from the "choosiness" of females. It is only with the introduction of male intrasexual selection that gender differences in displayed attitudes towards risk can be extended to non-mating contexts. However, the appeal to male intrasexual selection does not in fact succeed in bridging the evolutionary biological argument and the economic conclusion about displayed attitudes towards risk in non-mating contexts.

To see this, recall that the idea behind the appeal to intrasexual (male) competition is that females engage in short-term mating to obtain offspring with good genes (perhaps among other things—Mays & Hill, 2004). However, given this, it would be in the males' biological interest to engage in risky behavior only to the extent that doing so helps them display their health, strength, and mental abilities (all of which are generally taken to have at least some genetic basis—Gangestad & Simpson, 2000). However, if "risky behavior" is understood in the way common in economics—namely, as concerning the choice of actions with mean-preserving spreads—then this antecedent is unlikely to be satisfied. There is no reason to take a disposition towards the choice of actions with mean-preserving spreads as evidence for someone having "good genes." This disposition does not say anything positive about the health, attractiveness, intelligence, etc. of the male in question. In fact, as noted in Chapter 2, if anything, the fitness of risky behaviors in this sense is *lower* than that of less risky behaviors. In general, fitness is a *negative* function of the variance of the decision options (Gillespie, 1977).

In short, on the standard economic measures of risk, the previous evolutionary biological argument does not provide any reason whatsoever to think that males display more risk-love than females *apart from when it comes to mating decisions*; if anything, the opposite is true. Hence, this argument fails to connect to the economic phenomena to be explained in the appropriate manner (even taking for granted the—at least debatable—assumptions of the existence of a link between the fitness of a decision option and its utility/prospects/regrets/simple heuristics, and the "null model" assumption of no differences across the genders unless reasons to the contrary are produced).

This conclusion changes if other senses of "risky behavior" are concerned. So, if "risky behavior" is understood as behavior that carries a relatively high risk of injury or death (Byrnes et al., 1999), then, by the classic arguments, for something like a handicap principle, male risk-love—across many different contexts—*can* be sexually selected for (Zahavi, 1977; Clutton-Brock, 2007; Skyrms, 2010). That is, if having a peacock's tail is considered risky (in the sense of making predation more likely or making it harder to obtain food), and if having such a tail, in virtue of its costliness, signals good biological endowments, then it may be adaptive for males to be more risk-loving that females. However, while interesting, this conclusion is not the one that is of interest in the present context: it just does not speak to the kind of phenomena economists are concerned with when discussing attitudes towards risk.[25] For example, this argument does not address the question of what kinds of investment decisions are taken in the stock market. Investment decisions of this kind do not constitute flashy, dangerous behavior, but anonymous bets on more or less variable options. The latter is the main issue of interest here, though—and hence, appealing to other senses of "risky" behavior is not helpful in this context.[26]

All in all, where does this leave the argument? The main conclusion that needs to be drawn here is that, *in the present context*, the argument lacks internal coherence. The argument may well provide evolutionary biological support for an evoked account of gender differences in *some*, mating-related, behavioral dispositions—though this depends on the plausibility of the null-model assumption and the assumption that fitness and utility/prospects/regrets/simple heuristics are correlated (both of which require further discussion). However, this argument does not provide evolutionary biological support for gender differences *in the kinds of behavioral dispositions at stake here*—namely, dispositions to choose or avoid high-variance options. There is thus a major disconnect between what the argument can establish and what we are trying to find out: the purported explanans does not match the explanandum. Hence, as far as this argument is concerned, outside of mating contexts, females and males may well have different displayed attitudes to risk for fundamental reasons; this argument cannot provide evolutionary biological evidence for thinking otherwise. Two final points concerning this argument need to be made.

First, this assessment of course does not mean that other evolutionary biological arguments could not be formulated that do make for evolutionary biological evidence for or against the evoked or fundamental account of gender differences in attitudes towards risk. (Indeed, it is intriguing to consider to what extent a gene-cultural argument parallel to that of the next sub-section could be formulated here as well—see e.g. Gneezy & Leonard, 2009.) The point here is just that, appearances to the contrary (perhaps), this evolutionary biological argument does not do so. This is an important conclusion not just because this argument is the central evolutionary biological argument in the literature, but also because it clearly illustrates the promises and challenges of this version of the evidential form of evolutionary economics. The promises lie in the fact that evolutionary biological evidence for some economic issues really is in the offing; the challenges lie in the fact that it is not easy to formulate internally coherent evolutionary biological arguments with the appropriate conclusions.

Second, as noted earlier, the data on gender differences in attitudes towards risk are not fully clear to begin with. (For example, these differences seem to be much weaker—or not present at all—if females make economic decisions in a peer-group of other females only: Booth et al., 2014; Booth & Nolen, 2012; J. A. Nelson, 2015; Byrnes et al., 1999.) Given this, if it were to turn out that there are no gender differences in attitudes towards risk in non-mating contexts after all, then that would obviously be consistent with the previous evolutionary biological argument. However, this consistency would then be merely trivial: it would rest entirely on the null-model presumption of there being no gender differences in attitudes towards risk to begin with. No further support to this presumption would be provided by the appeal to differences in minimal parental investment or to male intrasexual competition. Put differently, any reason for thinking that there are no gender differences in attitudes towards risk in non-mating contexts would then just be based on how the relevant empirical studies turn out—and the previous evolutionary biological evidence would fall by the wayside. This would only change if we had a positive reason for thinking that the null model assumption is true; however, the possible existence of such a reason merely reduces to the previous point that there may well be other evolutionary biological arguments that could, in the future, be formulated here. (The next subsection returns to issues related to this point as well.)

2. Cultural Differences in the Propensity to Share

When considering cross-cultural differences in the extent to which people are disposed towards sharing, it is again useful to begin by making clearer what the fundamental and evoked accounts come down to here. In this context, both of these accounts agree that something about the cultural environment of the different human groups in question is at the root of the differences—they just differ over what this is (see e.g. Henrich et al., 2005, pp. 841–846).

The evoked account suggests that underlying the diversity in the observed sharing dispositions is a fundamental universality. All cultures rely on the same *fundamental* sharing dispositions; it is just that these sharing dispositions contain variables or parameters that (a) have different values in different cultural settings, and/or (b) which are "instrumented"— estimated using available data—differently in different cultural settings. As a result of (a) and (b), there will be different outcomes as far as the resultant sharing behavior is concerned, despite the sharing dispositions themselves being fundamentally the same (see e.g. Kenrick & Sundie, 2005). So, it may be that, in all cultures, people are more inclined to share with biological kin than with non-kin (Kenrick & Sundie, 2005; Cronk & Gerkey, 2007; Gowdy et al., 2013). However, it may also be that, in different cultures, (a) people are differentially likely to interact with kin, and (b) estimating who is biological kin is differentially difficult (Cronk & Gerkey, 2007; Markman et al., 2005). For example, if all the children in the community are raised together and separately from the parents (as in a Kibbutz), or if the culture is polyandrous and children grow up with their mothers, then determining who is related to who and to what extent can be complex and needs to rely on more or less reliable cues. In this way, there can be major differences in the *apparent* sharing dispositions of people from different cultures—even if, *fundamentally*, their sharing psychologies are identical (Hintze & Hertwig, 2016; Markman et al., 2005).[27] Henrich (2000, p. 973) expresses this quite clearly:

> Like most efforts to model human behavior in economics, these new approaches, implicitly or explicitly, make certain universalist or panhuman assumptions about the nature of human economic reasoning. That is, they assume that humans everywhere deploy the same cognitive machinery for making economic decisions and, consequently, will respond similarly when faced with comparable economic circumstances.

By contrast, the fundamental account suggests that sharing dispositions, even at a fundamental level, are learned from the cultural environment (Henrich et al., 2005, pp. 812–814, 842–846; Henrich, 2000,
p. 973). The fundamental account may agree that there are some biological constraints on the nature of the sharing dispositions that are being
transmitted and adopted; however, much of this transmission and adoption is influenced by factors unique to specific cultures. This is parallel
to other cultural phenomena, such as reproductive norms (who gets to
reproduce with who). There, too, biological constraints exist, but there,
too, cultural factors loom large in their explanation (Henrich, 2015;
Henrich & McElreath, 2007; Boyd & Richerson, 2005). In turn, the cultural factors that can lead to a divergence in sharing dispositions are
myriad. They could comprise accidental differences in the models that
are being copied (e.g. whether a more generous or a less generous individual happens to be identified as a good model), differences in the learning mechanisms involved (e.g. learning from few models vs. learning
from a majority of models), as well as correlations with other features
of the relevant culture (how much use it makes of market interactions,
how stratified it is, or how warlike it is), each of which further rests
on other gene-cultural factors (Henrich, 2015; Henrich & McElreath,
2007; Boyd & Richerson, 2005; Henrich et al., 2005). In this way, the
fundamental account suggests that differences in the sharing behavior of
people from different cultures are largely due to the fact that people are
simply differentially disposed to share with others—even holding fixed
the value and kind of the resource to be shared, as well as the nature
of the relationship between the sharer and the recipient. Again, Henrich
(2000, p. 973) expresses this quite clearly: "notions about what is fair
and/or what deserves punishment are culturally variable."

Which of these two accounts is more plausible? Assessing this is difficult, as both of these accounts seem, in many ways, equally internally and
externally coherent. On the one hand, according to *both* of the accounts,
differences in human economic decision-making with regards to sharing do *not* have genetic underpinnings. This is plausible, as few (if any)
systematic genetic differences across cultures have been found (Boyd &
Richerson, 2005; Kenrick & Sundie, 2005; Galanter et al., 2017).

On the other hand, both of the two accounts are equally consistent
with widely accepted evolutionary biological models of sharing. In particular, it is now widely accepted that giving away resources can be highly
adaptively advantageous—but only in the right circumstances (Gardner
et al., 2011; A. S. Griffin & West, 2002; West et al., 2011; West et al.,
2007, 2008; Sober & Wilson, 1998; Okasha, 2006; Birch & Okasha,
2014; Bowles & Gintis, 2011). For example, sharing can be adaptive
if it is directed at kin, in-group members, or if it leads the recipient to
reciprocate in the future. Note, though, that none of these claims hold by

necessity. Whether sharing with kin, for example, is adaptive depends on the benefit the sharing yields to the recipient, the cost to the sharer, the degree of relatedness between sharer and the recipient, the probability that the recipient will reciprocate the sharing, the population structure of the organism in question, etc. (see also Schulz, 2018a). Importantly, both the fundamental and the evoked account are able to do justice to these facts.

This is relatively obvious when it comes to the evoked account, as the account was devised with this in mind. According to this account, all humans share in line with the same, biologically evolved sharing dispositions—it is just that using and instrumenting these dispositions in different cultural settings leads to differences in their superficial sharing dispositions (Kenrick et al., 2008; Kenrick & Sundie, 2005; see also Cosmides & Tooby, 1992).[28]

However, the fundamental account can accommodate the biological constraints on the evolution of sharing dispositions just as well. So, as also just noted, the fundamental account allows for—and to some extent is based on—an interaction between biological and cultural pressures. While, according to this account, sharing dispositions are culturally learned, this does not mean that there are no biological influences on which sharing dispositions are more easily acquired, nor does it mean that the adoption of different sharing dispositions does not have differential adaptive consequences (Boyd & Richerson, 2005; Henrich, 2015; Henrich & McElreath, 2007; Bowles & Gintis, 2011). For an analogy, consider the diversity in politeness norms. There is every reason to think that politeness norms are culturally learned, but there is also reason to think that politeness norms that fit to our evolved emotional endowments (e.g. in terms of disgust) are more likely to persist and spread (Nichols, 2004). Similarly, while reproductive norms also seem to spread culturally, this spread is very plausibly curtailed by what is biologically advantageous. The spread of celibacy, for example, is plausibly being curtailed by its biological disadvantageousness (Boyd & Richerson, 2005). In this way, we can see the cultural learning of sharing dispositions as "filling out" the space left open by the biological constraints.

For these reasons, it should be concluded that both accounts—the fundamental and the evoked—have the resources to account for all the empirical sharing behavior observed. The fundamental account appeals to the gene-culture coevolution of sharing dispositions, and the evoked account to the fact that people live in different circumstances (especially as far as their propensity to interact with kin is concerned) and/or instrument relatedness differently in different cultures.[29] Furthermore, both of these accounts are consistent with the fact that there are few genetic differences across cultures. Hence, so far, both accounts seem about equally externally and internally coherent. However, it turns out that there is a reason to think that the fundamental account ultimately comes out on top here. This reason centers on the fact that this account is more coherent with certain *methodological* norms of theory choice than the evoked account.

To see this, it is best to begin by returning to the core differential prediction of the two accounts. This prediction centers on how people share with kin relative to non-kin. The evoked account—as just noted—predicts that they are more likely to share with biological kin (holding the value and cost of the resource to be shared constant), regardless of their cultural background. Biological kin are more likely to share genes, so genetic dispositions for sharing can spread if they focus on kin (Hamilton, 1964; Gardner et al., 2011). Of course—as also just noted—this is consistent with there being some cultural differences in how people share with kin, since the "instrumentation" of kin—i.e. how biological kin are detected and conceived of by people—can differ across cultures (Cronk & Gerkey, 2007).

However—and this is the key point for present purposes—on the evoked account, these cultural differences in dispositions to share with kin are not arbitrary, but likely to closely track the biological facts of kinship. "Instrumentations" of who is kin with who are important in a wide range of decisions beyond decisions about who to share with: for example, estimates of who is related to whom are crucially important for mating decisions (e.g. to avoid mating with close relatives). For this reason, there is evolutionary pressure on these "instrumentations" being good estimates of who is *actually* biologically related to whom. This matters, as it suggests that the evoked account predicts that there will be relatively little cultural variation in terms of how people share with kin: controlling for (perceived) differences in kinship, all cultures should show about equal levels of sharing (Kenrick & Sundie, 2005).

By contrast, it is a core feature of the fundamental account that the cultural evolution of sharing dispositions—while (potentially) checked and constrained by biological factors—operates quite independently of biological features (Boyd et al., 2011; Henrich, 2015; Henrich et al., 2005; Henrich & McElreath, 2007). In turn, this predicts that this independence of biological constraints will show up somewhere (Henrich et al., 2001, p. 75). Indeed, this is true in other contexts. For example, the cultural differences in mating institutions—including the existence of celibacy in some contexts—is a clear strike in favor of mating dispositions partly evolving through cultural transmission. If human mating dispositions had evolved purely biologically, we would not expect to see this kind of variance in mating behaviors (Henrich, 2015; Boyd & Richerson, 2005). Much the same is true for sharing dispositions. If sharing behavior evolves partly culturally, we should expect there to be significant cultural variability *even if* the sharing concerns biological kin. While it may not always be biologically adaptive to share less with close kin than with more extended kin (or even non-kin), the independence of cultural pressures and biological pressures suggest that this will occur relatively often.

In this way, cross-cultural sharing behavior towards kin can be seen as a litmus test for which of the two theories of the nature of human sharing dispositions is more compelling. The core difference between these two accounts—whether there is cultural variability in the nature of human sharing dispositions, or merely variability in the circumstances of the use of the same human sharing dispositions—comes out particularly clearly when it comes to sharing with kin. Three further points about these differential predictions need to be noted.

First, the differential predictions of the two accounts are, as just noted, relative not absolute. The fundamental account does not predict that there will not even be a tendency across cultures for people to be more inclined to share with kin, and the evoked account does not predict that in no culture will people not be more inclined to share with kin than with non-kin. These differential predictions are matters of degree, and—so it must be acknowledged—therefore less clear-cut and precise. In turn, this means that only a fairly systematic investigation of cross-cultural sharing dispositions with regards to kin can fully resolve this issue; short of this, only hints of an answer can be given.

For now, though, the hints that are available suggest that the fundamental account is more plausible. On the one hand, given that the evoked account predicts that people across cultures should be more inclined to share with biological kin (ceteris paribus), it should be the case that more close-knit, family-based cultures should show higher degrees of sharing in the ultimatum game. In these cultures, most sharing happens among kin, so that—given the evolutionary importance of accurate assessments of kinship—people should be expected to work with the heuristic that sharing even with an anonymous member of their culture involves sharing with relatively close kin (Kenrick & Sundie, 2005). However, this is not what we find. There is much variation in the sharing dispositions among close-knit, family-based cultures, with some cultures showing very little sharing behavior (Henrich et al., 2005). This thus speaks against the prediction of the evoked account.

On the other hand, studies of cooperation and sharing have shown that kinship is less relevant to cooperative decisions in a number of cultures than cultural factors, such as "lineage-membership" (in matrilineal or patrilineal societies) (Alvard, 2003; Cronk & Gerkey, 2007). For example, in a study of the ultimatum game among the Bwa Mawego (Macfarlan & Quinlan, 2008), it was found that "average relatedness to the village did not predict the size of proposals" (Macfarlan & Quinlan, 2008, p. 304), and that "[t]he matrifocal nature of village life may pattern the altruism received across the life course such that males and females develop different sets of models of fairness conditional on family characteristics" (Macfarlan & Quinlan, 2008, p. 306). Similarly, whale-hunting crews among the Lamalera—which need to be highly cooperative—are formed, in the first instance, of members of the same lineage (not kin). While this latter study, and some others like it (Cronk & Gerkey, 2007), do not consider the ultimatum game directly, they at least provide reasons to think that sharing is not always kin-driven.

For this reason—and though there are of course also cases that do fit the predictions of the evoked account quite well (see e.g. Thomas et al., 2018)—the most plausible conclusion to draw from this research at this point is that people are not *per se* more inclined to share with kin than with non-kin. Rather, their dispositions to share seem heavily influenced by cultural factors such as the lineal organization of the culture in question. Of course, this conclusion can change as further studies are being made (see also Macfarlan & Quinlan, 2008, p. 308). As matters stand, though, the most reasonable inference to draw here is that there is some predictive support for the fundamental account.

The second point to note is that what is at stake in the present context is differential *predictive* success, not differential ability to *accommodate* the data. As noted earlier, as far as *consistency* with the data is concerned, both accounts are equally strong: they have enough free parameters to make sense of any given empirical finding. The point made here though is that there are differences in whether the accounts can *predict* all the data equally well. This matters, since—as also noted in the previous chapter—there is a widely accepted methodological principle of favoring predictive success over accommodative success: ceteris paribus, it is reasonable to favor more predictively accurate theories to ones that are less so (Hitchcock & Sober, 2004; White, 2003). Predicted fit to the data speaks more in favor of the theory providing an accurate representation of the processes actually going on in nature than non-predicted fit. There is no need to adjust the theory to make it fit to the data, but the theory achieves that fit automatically through the way it is structured (White, 2003; Hitchcock & Sober, 2004).

Third and relatedly, note that the differences in the predictive accuracy of the two accounts become visible only if the *evolutionary foundations* of these accounts are considered. The mere fact that the evoked account locates the source of the variability in sharing behaviors on different instrumentations of the same underlying sharing dispositions, rather than in fundamentally different sharing dispositions, would not lead us to see the former account as predicting less cultural variability than the latter one. Rather, the predictive asymmetry becomes visible only once the facts are considered that (a) "instrumentations" face evolutionary pressures to be accurate and (b) the gene-culture coevolution of sharing dispositions involves two *independent* evolutionary pressures. It is points (a) and (b) that underlie the predictive differences of the two accounts. Point (a) grounds the fact that, on the evoked account, we should expect people to share more with kin than with non-kin (ceteris paribus), and point

(b) grounds the fact that, on the fundamental account, sharing with kin should not be expected to be drastically more important to people than sharing with culturally significant non-kin. In this way, it becomes clear that the appeal to evolutionary biology does important argumentative work here: without this appeal, we would not be able to obtain the present predictively focused reason to favor one of the accounts over the other.

Overall, therefore, we arrive at opposite conclusions concerning the two differences in economic decision-making. Whether or not gender differences in attitudes towards risk are evoked or fundamental is not currently well illuminated by an appeal to evolutionary biological considerations. By contrast, evolutionary biological considerations *do* help in bringing out that cultural differences in sharing dispositions are likely to be due to the gene-cultural evolution of fundamentally different sharing dispositions in different cultures. In this way, the discussion of these two cases also illustrates the importance of paying attention to the internal and external coherence of evidentially evolutionary economic arguments: providing compelling (partial) evolutionary biological accounts of some economic phenomenon is not as easy as is sometimes claimed (Richardson, 2007).

V. Conclusions

This chapter has focused on differences in human economic decision-making. In particular, there seem to be gender differences in attitudes towards risk and cultural differences in sharing dispositions. What is not clear is whether these differences are merely evoked—the result of a common (pluralistic) decision-making machinery that is merely embedded in different environments—or fundamental—the result of different (though overlapping and pluralistic) decision-making machineries.

I have argued that while determining when either of these explanations is the most plausible one is partly a psychological or anthropological matter, it is also a question that evolutionary biological considerations may be able to help resolve. In particular, while most likely unable to *fully* determine whether evoked or fundamental explanations of a case of human diversity in economic decision-making are more compelling, evolutionary biological considerations can—at least in principle—play an evidential role in this determination.

I then looked at two specific arguments in this context, and came to two conclusions. First, evolutionary biological considerations are *not* (currently) useful for advancing the debate concerning the nature of gender differences in attitudes towards risk. Second, by contrast, evolutionary biological considerations *can* be used to support the hypothesis that cultural differences in sharing dispositions are fundamental in nature. A major methodological lesson of this discussion is that the plausibility of an evolutionary biological argument depends on its internal and external coherence. For this reason, the overarching upshot here is again qualified: also when restricted to what is known about *human* evolutionary history specifically, an appeal to evolutionary biology, can—but need not—make new evidence available for the resolution of some economic questions.

Suggested Further Reading

Ashraf and Galor (2013) is a much-discussed take on using evolutionary biology to explain differences in economic development. Gneezy and Leonard (2009) is an anthropologically based study of differences in competitiveness in a matrilineal and a patriarchal society. Blake et al. (2015) is a developmental psychological look at the development of fairness norms in different cultures. D.M. Buss and Hawley (2010) collects a variety of papers about the evolution of individual differences in psychological traits. Boyer and Petersen (2017) present evolutionary biological reasons for some of the economic beliefs commonly held by people. Baumard (in press) develops a life-history-theory-driven argument for why the British industrial revolution occurred where and when it occurred. Saad (in press, 2013) are evolutionary psychology-based treatments of consumer behavior.

Notes

1. As in the previous chapter, no assumption is made here whether economic decision making is distinct from decision making more generally. All the conclusions of this chapter can be formulated with either a narrow or a broad account of the latter in the background.
2. As noted in the previous chapter, this focus may also be useful for clarifying the uniqueness of human economic decision making compared to animal economic decision making.
3. A complication here is that many different characterizations of risk are used in these studies (Eckel & Grossman, 2008; Byrnes et al., 1999), which makes the assessment of these studies somewhat tricky. I return to this point momentarily.
4. Of course, the same distinction can be and has been made concerning the explanation of differences in psychological traits more generally: see e.g. Tooby and Cosmides (1992) and Boyd and Richerson (2005).
5. As will also become clearer momentarily, I use the term "evoked" in a slightly more restricted sense than is common in the literature.
6. A more complex variant of this explanation concerns cases of niche construction (Odling-Smee et al., 2003). In these cases, the different selection pressures that different populations of organisms have faced are at least partly their own doing: these different populations have altered their environments in different ways, thus leading to different selection pressures. See Chapter 2 for more this.
7. A traditional concern for this explanatory strategy is that it is commonly thought that evolution by natural selection is a fairly slow process (see e.g. Gillespie, 1998). However, this concern should not be overstated. First, there is now a consensus that evolution by natural selection can operate

at relatively fast rates (see e.g. Hawks et al., 2007). Second, there is some reason to think that human populations have, at least sometimes, occupied different selective environments for sufficiently long periods of time (see e.g. Beall et al., 2010; Huerta-Sanchez et al., 2014; D. M. Buss & Schmitt, 1993; Hawks et al., 2007). For these reasons, this is an explanatory strategy that in general does need to be taken seriously.

8. Note that this assumes that organisms cannot (probabilistically) play both hawk and dove. In the latter case, it is possible that this situation leads to human homogeneity, in the sense that all humans play the same mixed strategy of hawk and dove (see also Orzack & Sober, 1994).

9. Of course, one could align this case with the previous one concerning different selection pressures by seeing hawks facing different numbers of doves as occupying different selective environments. Either way, this is an important possible explanation of human diversity.

10. Note that it is an open question whether this explanation needs to appeal to something like group selection—and whether this would be problematic (for discussion, see e.g. West et al., 2007, 2008; D. S. Wilson, 2008).

11. Note that the same point about the speed of genetic evolution as raised in the case of directional selection can apply here as well.

12. A variant of this kind of case appeals to the fact that different groups of humans might systematically live in different environments that in combination with their (shared) psychology for learning from these environments can lead to systematic differences in human ways of being or acting. An especially intriguing version of this sort of scenario concerns cases where environments are specifically constructed so as to facilitate learning of a certain sort (Sterelny, 2012a, calls this "apprentice learning"). This then is a cultural analogue to niche construction and can make for *heritable* differences among different human beings (see note 6). While there are interesting differences to the classic cases of gene-culture coevolution sketched in the text, for present purposes, these two sets of cases will be grouped together.

13. On strongly embedded views of organismal development (Griffiths & Gray, 1994; Griffiths & Stotz, 2018), this distinction cannot be drawn. However, a related distinction among different ways the corn plants may develop still is. See also the following.

14. Similarly, it may be that the difference in whether someone is more politically conservative or more politically liberal is triggered by the (subconsciously) detected amount of pathogens in the environment, perhaps at a key point in their youth (Thornhill et al., 2009). While this would imply that underlying the diversity in how liberal different humans are is a common conditional disposition to react to the amounts of pathogens around, this shared basis does not call into question the fact that from the point of view of political science it is useful to divide voters into "liberals" and "conservatives." Note also that as such this explanatory strategy leaves it open *why* there are the genetic, developmental, or behavioral switches that bring about human differences—whether they are accidental byproducts of the way humans develop and act, or whether they have been specifically selected for. What is at the core of this account is just the idea that human diversity is not to be seen as a *relatively* "deep" feature of human biology, but merely as something that is the result of much human homogeneity.

15. See e.g. Wessinger and Rausher (2014) for an illustration of why taking these sorts of issues into account in evolutionary biological inquiries can be important.

16. For a similar defense of the appeal to evolutionary biological considerations in psychology, see Schulz (2018b).

17. As will be made clearer in the next two chapters, and as has been noted in Chapter 1, if the appeal to evolutionary biological considerations is merely done in a heuristic way, these constraints do not apply as such (though other constraints do). The point here is just that these constraints hold for *evidential* forms of evolutionary economics.

18. Some aspects of this argument are more hinted at than explicitly stated. Still, these hints are clear enough (even if scattered through a number of different publications) that it is relatively straightforward to develop them into the explicit argument laid out here.

19. Differences in minimal parental investment have been widely used—and for a long time—in behavioral ecology to make sense of various aspects of non-human mating systems and behaviors (Trivers, 1972; Alcock, 2013).

20. Much the same distinction has come to be recognized in the context of non-human animals as well: see e.g. Cockburn (2013) and Dunn and Cockburn (1999).

21. It is worthwhile to note that several scholars have pointed out that there are some costs to short-term matings for males as well. In particular, they may obtain sexually transmitted diseases. Depending on the details of the case, this can complicate the evolution of attitudes towards risk further (Boots & Knell, 2002; Thrall et al., 2000; Hanna et al., 2002). However, for present purposes, these complications need not be addressed.

22. This also implies that the prospect theoretic reference point of females is likely to be shifted relative to males; for females, more short-term mating decisions involve potential losses than for males. Given that prospects are assumed to be downweighted in the loss-dimension relative to the gain-dimension—i.e. that people are generally loss-averse—this implies that females will likely to be more loss-averse than males when it comes to short-term mating decisions (McDermott et al., 2008).

23. This evolutionary biological account for an evoked explanation of gender differences in attitudes towards risk is underwritten by two further considerations. First, Fessler et al. (2004)—for reasons that parallel the arguments just laid out—suggest that human males may be more prone to experiencing anger, and human females to experiencing disgust. Anger is an emotion that if adaptively triggered leads to adaptive if daring actions, and disgust is an emotion that if adaptively triggered leads to the avoidance of maladaptive situations. Hence, this too, will lead to a gender difference in risk aversion. Second, Sapienza et al. (2009) have shown that at least for an intermediate range, increases in testosterone in males lead to increases in risk aversion. Given the known role of testosterone in mediating male social behavior (Montoya et al., 2012), this thus further supports the previous argument.

24. It may be that these differences sometimes become particularly pronounced when the participants are in a "mating frame of mind" (Y. J. Li et al., 2012). However, it is not clear that this holds for all such differences in displayed attitudes towards risk (Byrnes et al., 1999), and this is anyway still different from concerning mating decisions per se. See also note 26.

25. It is also noteworthy that the "handicap principle" is not so straightforwardly true (Skyrms, 2010).

26. For the same reason, the argument cannot even establish the conclusion that there will be gender differences in displayed attitudes toward risk if the relevant agents are *in a mating context*—i.e. have been primed to think about possible mating opportunities, but do not make decisions directly about who to mate with (as in the research of Y. J. Li et al., 2012). The argument really just speaks to *mating decisions per se*—i.e. decisions about who to mate with.

27. Another example of this sort of case has been provided by Henrich et al. (2005, p. 811). They suggest that the fact that even hyper-fair offers in the ultimatum game are frequently rejected among the Au and Gnau is because in this culture the acceptance of a gift is taken to imply an obligation to repay this gift at a later date. If so, then this kind of rejection of hyper-fair offers should not be seen to display a different attitude towards resource division—it is just an aspect of the fact that in this culture gift giving is a much more dynamically extended affair that includes repayment of the gift later on. If this point is taken into account, the differences in the sharing dispositions between this culture and others might well disappear: *holding the value of a gift fixed* (which may include considering any obligations to repay the gift later), people from different cultures may display the same sharing dispositions (Kenrick & Sundie, 2005).

28. This is also worth noting in regards to the remarks of Henrich et al. (2005, p. 844) concerning the ultimatum game: "[i]f our fitness maximizer is the proposer and she knew that the respondent is also a fitness maximizer, she should offer the smallest positive amount possible, knowing that the fitness-maximizing respondent will accept any positive offer. This simple fitness maximizing prediction is not supported in any society. Thus, our work [i.e. the cross-cultural findings sketched previously] provides an empirical challenge to evolutionary theory." This does not follow as long as by "fitness-maximizing" is meant *inclusive fitness* maximization—or a related notion, such as neighborhood-modulated fitness (Hamilton, 1964; Gardner et al., 2011; Rubin, 2018; Grafen, 2006; Sober & Wilson, 1998; Godfrey-Smith, 2008). It can make *purely biological* sense to share resources with others (though not in all certain circumstances). See also Piccinini and Schulz (2019) for more on this.

29. This is important to note, as it is not always given the attention it deserves. So, Henrich (2000, p. 973) says "if the Machiguenga results [and others like them] stand the test of scrutiny and can be replicated elsewhere, then the assumption that humans share the same economic decision-making processes must be reconsidered." However, this is not the case—as long as the evoked approach is understood properly.

6 Equilibrium Modeling
Economics, Ecology, and Evolution (The Heuristic Project)

I. Introduction

A traditional evolutionary economic criticism of mainstream economic analysis is that the latter is too strongly focused on equilibrium models and thus fails to do justice to the complex and dynamic nature of real economic systems.[1] R. Nelson and Winter (2002, p. 24) express this point as follows:

> In recent years, evolutionary arguments have begun to come back into economics, at least around the fringes of the field. This change is partly the result of a growing awareness that standard neoclassical theory cannot deal adequately with the disequilibrium dynamics involved in the kind of competition one observes in industries like computers or pharmaceuticals or, more broadly, with the processes of economic growth driven by technological change.

Similarly, Carlaw and Lipsey (2012, pp. 736–737) note:

> Economists face two conflicting visions of the market economy, visions that reflect two distinct paradigms, the Newtonian and the Darwinian. In the former, the behaviour of the economy is seen as the result of an equilibrium reached by the operation of opposing forces—such as market demanders and suppliers or competing oligopolists—that operate in markets characterised by negative feedback that returns the economy to its static equilibrium or its equilibrium growth path. In the latter, the behaviour of the economy is seen as the result of many different forces—especially technological changes—that evolve endogenously over time, that are subject to many exogenous shocks, and that often operate in markets subject to positive feedback and in which agents operate under conditions of genuine uncertainty. . . . In this paper, we consider, and cast doubts on, the stationarity properties of models in the Newtonian tradition. These doubts, if sustained, have important implications for understanding virtually all aspects of macroeconomics, including of long-term economic growth, shorter term business cycles, and stabilisation policy.

Similar remarks can be found in a number of other sources in the literature (see e.g. Vromen, 2001; Young, 1998; Groenewegen & Vromen, 1999; Hodgson, 1999, pp. 258–260; Lawson, 2006; Galla & Farmer, 2013; Berger, 2009).

The aim of this chapter is to assess the plausibility of this criticism further. More specifically, the goal in what follows is both to determine whether it is *true* that the heavy reliance on equilibrium models in economics is problematic, and whether and how an appeal to *evolutionary biology* can prove useful towards answering this question (positively or negatively). In line with the rest of this book, therefore, the approach here is simultaneously substantive and methodological: it tries to both make a contribution to a contested issue in economics, and assess whether and how an appeal to considerations from evolutionary biology is helpful in making this contribution.

To achieve this, I consider the discussion in evolutionary ecology surrounding the extent to which ecosystems can be expected to be stable, and analyze whether, when, and how insights from that discussion can be translated into the economic case. The upshot of this analysis will be the suggestion—countering these evolutionary economic claims—that, in many (but not all) cases, economic systems *will* be well analyzable with equilibrium models. In turn, this is due to the fact that in many (but not all) cases it is plausible that they are "sorted" systems (in a sense to be made precise momentarily). Importantly, this suggestion will be derived from the idea that many ecosystems, too, are well analyzable with equilibrium models because they are "sorted" systems. However, I also show that the ways in which ecosystems and economic systems are sorted systems is very different. For this reason (among others), I further make clear that whatever usefulness the appeal to evolutionary biology has in this context, it is only *heuristic* in nature. In this way, the present discussion can also be used to make clearer the nature of cross-disciplinary heuristic support more generally.

The chapter is structured as follows. In section II, I precisely define how I understand equilibrium models. In section III, I make the criticism surrounding the use of equilibrium models in economics clearer. In section IV, I summarize key lessons from the debate surrounding the stability of ecosystems in evolutionary ecology. In section V, I assess to what extent these lessons can be applied to the debate surrounding the use of equilibrium models in economics. I conclude in section VI.

II. Equilibrium Models

Equilibrium models, as they are understood here, are models of target systems that, eventually, settle into some kind of stable state, and where the nature of this stable state is the focus of the analysis of the relevant system. To refine this understanding of equilibrium models, it is best to begin by making three clarificatory points.

First, I here leave it open what a *model*, exactly, is. As far as the discussion of this chapter is concerned, (equilibrium) models could be set-theoretic structures, sets of equations, computer simulations, pictures/graphs, or even physical objects (for discussion of this, see e.g. Morgan, 2012; Morgan & Morrison, 1999; Da Costa & French, 2003; Weisberg, 2013; Maki, 2009; Vorms, 2011; Toon, 2012). While the discussion to follow focuses on mathematically based models, this is done only for expositional convenience; the conclusions to follow also hold for other types of models. Similarly, and for the same reasons, the relationship between models and theories does not need to be further discussed here (Morgan & Morrison, 1999; Frigg & Nguyen, 2016).

Second, in what follows, I presume that equilibrium models in economics aim to represent features of the real world. That is, I presume that it is the goal of these models to describe and provide understanding about some causal system in the world—they aim to capture how the system behaves at different times (see also Potochnik, 2010; Sober, 1983). The reason behind the presumption of a representational aim of economic modeling is that the idea that all economic models are just tools for predicting economic phenomena without any kind of representational link to the world (Friedman, 1953) has widely come to be seen as implausible (see e.g. Hausman, 2008). While it may be true that *some* economic models are purely predictive in nature, it is hard to do justice to all aspects of work in economics in this manner (Hausman, 2012, chap. 3; but see also Reiss, 2012). However, no commitment is made here to any kind of strong form of scientific realism (see e.g. Massimi, 2018). All that is presumed in what follows is that equilibrium modeling in economics cannot be defended solely in terms of the fact that—for some unknown reason—it yields successful predictions. Put the other way around, the nature of economic reality is taken to be an important constraint on the plausibility of an economic model. This is consistent with the nature of that constraint not being straightforward (Carlaw & Lipsey, 2012).[2] (I return to some aspects of this point in the next section.)

Third and relatedly, exactly how this representational nature of economic models comes about is something I leave open. Again, there is much discussion of this point (see Frigg & Nguyen, 2016, for an overview). Perhaps models represent their targets by being similar to them in some way (Perini, 2010; Giere, 2004), or perhaps they do so by being structurally isomorphic to the latter (van Frassen, 2008), or perhaps the representational nature of a model reduces to the fact that it allows researchers to make appropriate and useful inferences (Suarez, 2004; Contessa, 2007). Similarly, I here leave open how the fact that many (equilibrium) models are idealized contributes to their ability to represent their target phenomena (Potochnik, 2017; Batterman, 2009; Morrison, 2015; Weisberg, 2007). Finally, I leave open exactly what features of the real world are being represented by a model: whether it is causal capacities (Cartwright,

1999, 1989), modal facts (Sugden, 2000), or something else altogether (Hausman, 1992, chap. 5; Maki, 2009). The rest of the discussion of this chapter is consistent with any (generally satisfactory) characterization of how the representational link to the world beyond the model is obtained, and this link will therefore not be further specified.

With these three points in the background, it becomes possible to spell out equilibrium models in more detail. As noted earlier, at the core of these models is the fact that they have stable states that form the fulcrum on which the analysis of their target system rests. These stable states, in turn, are those states of the model system that it converges to if given sufficient time and to which it returns if disturbed (for similar understandings, see e.g. L. Samuelson, 2002, pp. 57–58; Carlaw & Lipsey, 2012). Note that what counts as a "state" will depend on the system in question: it could be a static number (such as the number of firms in an industry), but it could also be a dynamic (such as the growth rate of the firms an industry). There are two further aspects of this characterization that need to be made more precise, however.

On the one hand, there is the question of how much time is said to be "sufficient" for the system to reach its stable states.[3] The less time the system is given—i.e. the faster it is required to approach the stable state—the more restrictive the definition of equilibrium is. By contrast, the more liberal the definition is—i.e. the more time the system is given to reach its stable state—the more actual systems it can be applied to, and even systems that take a long time to approach their stable states are addressed by the model. Fortunately, for present purposes, a detailed statement of the appropriate length of time is not necessary. Here, it can just be assumed that *some* (finite) appropriate value for this variable can be set (at least for the application of the model at stake).

On the other hand, it is important to note that different senses of the stability of a system's states can be distinguished. One of the strongest senses of stability is one that requires a system to return to a given stable state from *any* (relevant) starting position. A basic example of such an equilibrium is a variable $X_t \in \mathbf{N}$ that undergoes stepwise changes according to the equation $x_{t+1} = x_t + \dfrac{x_E - x_t}{|x_E - x_t|}$ (unless $x_E = x_t$, in which case $x_{t+1} = x_t$), with x_E an exogenously defined parameter.[4] In this case, the system will return to its equilibrium state x_E no matter where it starts out, or to where it gets perturbed. A slightly weaker form of equilibrium stability is one where the system has multiple equilibria with connected basins of attraction.[5] So, consider a variable $y_t \in \mathbf{N}$ that changes according to the equations $y_{t+1} = y_t + \dfrac{y_{EA} - y}{|y_{EA} - y|}$ if $y_t < 100$, $y_{t+1} = y_t + \dfrac{y_{EB} - y}{|y_{EB} - y|}$ if $y_t > 100$, and $y_{t+1} = y_t$ if $y_t = 100$ or $y_t = y_{EA}$ or $y_t = y_{EB}$ (with $y_{EA} < 100$ and $y_{EB} > 100$ exogenously defined). The system will return to a given stable state only if it is perturbed sufficiently little; stronger perturbations lead the systems to approach a different stable state. (Put differently, the system will approach some stable state from anywhere within the system— just not the same one.) A graphical example of this looks like this:

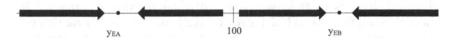

y_{EA} 100 y_{EB}

Figure 6.1 Example of multiple equilibria with their own basins of attraction

Another weakened conception of stability (which can also be combined with either single or multiple equilibrium settings) is one where the system does not return to a given state, but rather returns to a point close to the stable state.[6] (A different way of putting this is that, on this form of stability, perturbations are *bounded*.) In this case, therefore, the equilibrium is more coarsely defined than on the ones in the previous paragraphs: it is a set of states of the system, rather than just a single state.[7] Graphically:

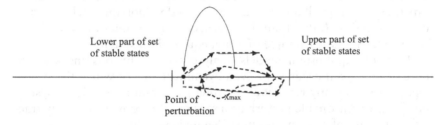

Lower part of set of stable states   Upper part of set of stable states

Point of perturbation x_{max}

Figure 6.2 Graph of a coarse-grained equilibrium around point x_{max}

Finally, the limiting sense of stability is provided by systems that contain (sets of) states that are such that the system will remain in them unless disturbed. In the rest of the discussion, though, these unstable equilibria will not play a major role, so a further discussion of them is not necessary here.[8]

While there is much that can be said about these different stability concepts (see e.g. Justus, 2008a), for present purposes, this is not necessary. Specifically, here it is sufficient to only presume that equilibrium models are models that approach some stable state from any state the system can reasonably be taken to be in.[9] This stable state need not necessarily be the same one throughout the system, nor need that stable state be extremely finely individuated; a coarse equilibrium is sufficient. The goal of these presumptions is to opt for a minimal standard that is precise enough to distinguish equilibrium models from other models without prejudging important issues that, if anything, should be the outcome of the investigation into the reasonableness of economic modeling in economics, not its starting point. This minimal standard is being met with the previous presuppositions.[10]

In particular, this rules out systems—sometimes known as "complex" or "dynamic" systems (though see also what follows)—which exhibit chaotic behaviors in at least a subset of their relevant dynamics.[11] In these systems, there is no fixed (even coarsely individuated) state that the system must tend towards; rather, the system may be at any of an array of indefinitely many states, depending on subtle differences in initial conditions. Furthermore, the states are not states that the system will persist in—with the important exception of trivial states that amount to the collapse of the system. Precise characterizations of these kinds of systems are hard to state (see e.g. Werndl, 2009; Strevens, 2005, for some different attempts at doing so), but fortunately, for present purposes, an illustrative example is enough.

Consider a system where some discrete quantity $q_t \, \varepsilon \, Z$ undergoes a step-change at every time period, with the probability of this change being positive (e.g. from 3 to 4) being determined by the simple rule $P(q_{t+1} > q_t) = 0.5$. In a case like this, the system follows a random walk: depending on where the system starts, the probability of ending up at any given state ($q_t = 17$, say) will differ, and the system does not approach or return to any given state (other than the trivial, collapsing states of $+\infty$ and $-\infty$). This is thus a classic example of a non-equilibrium system.[12]

All in all, equilibrium models analyze systems by positing a set of more or less specific states (or even dynamics), one of which the system approaches from any reasonable starting place and to which the system returns after reasonable perturbations. They thus provide an easily tractable summary of a given system (L. Samuelson, 2002).[13]

III. Equilibrium Models in Economics: Some Criticisms

There is no question that the use of equilibrium models in economics is widespread—they are the major tool with which actual economic systems are approached (Hausman, 1992; Mäki, 2013; Reiss, 2013).[14] To illustrate this, consider the Solow growth model (for a clear introduction to and discussion of the model, see Jones, 2002). This model makes for a good example of the centrality of equilibrium modeling in economics, as it is among the standard models taught to students in the discipline; it thus serves as a paradigm of the kind of work that members of the profession are meant to produce (see e.g. Mas-Colell et al., 1996; Varian, 1992; O. Blanchard, 2007).

In Solow's model of GDP (Y) growth, levels of GDP are assumed to be determined by a three-input production function—capital (K), labor (L), and a labor-modifying technology (A)—with each of the inputs being characterized by decreasing returns to scale (i.e. multiplying one of the inputs by v will only yield an increase of w < v in GDP). Expressed in terms of GDP and capital per "technology-modified worker" ($\tilde{k}$ = **K/AL** and $\tilde{y}$ = Y/AL), this implies

(1) $\tilde{y} = f(\tilde{k})$, with $f(vx) < vf(x)$.

Further, labor and technology are assumed to grow at fixed rates n and g (so that $\dfrac{\dot{L}}{L} = $ n and $\dfrac{\dot{A}}{A} = $ g), and capital is assumed to accumulate as the difference between a fixed savings component (sY) and a fixed capital depreciation rate (dK):

(2) $\dot{\tilde{k}} = s\tilde{y} - (n + g + d)\tilde{k}$

Given this, sooner or later, $\tilde{k}$ will stop accumulating: it will be the case that $s\tilde{y} = (n + g + d)\tilde{k}$. At this point—the equilibrium of the model—GDP per head (y = Y/L = A$\tilde{y}$) will grow at a rate that is just determined by the rate of technological progress g, since $\tilde{y}$ (as a function of $\tilde{k}$) will stop increasing.

This matters, as it can be used to analyze the behavior of the economy at various time slices. For example, assume that, at time t_0, the economy is not at its equilibrium state $\tilde{k}^*$—for example, assume that $\tilde{k}_0 < \tilde{k}^*$. We can then predict that capital will accumulate—i.e. the economy will grow at a rate faster than n—until $\tilde{k}_0 = \tilde{k}^*$. Alternatively, if at t_0, there is a change in the tax regime that makes saving more attractive—so that the saving rate increases from s to s'—an economy at equilibrium will grow until a new equilibrium at $\tilde{k}^{*'}$ is reached. In both cases, there will be a temporary boost in growth that gradually falls back down to the equilibrium growth rate of g. Figures 6.3 and 6.4 illustrate these points.

For present purposes, the key point to note about this model is that its focal point is the equilibrium growth rate. The understanding of the growth dynamics of actual economies provided by this model is "filtered" through its equilibrium states (Jones, 2002). This is a key feature of many other economic models as well: understanding of the economic system in question is provided by an analysis of the relevant model's equilibrium states and their perturbation dynamics.

However, as noted in the introduction, this kind of reliance on equilibrium models in economics has its detractors. Arthur (2015, pp. 3–4) provides a very succinct expression of this point:[15]

[The] equilibrium shortcut was a natural way to examine patterns in the economy and render them open to mathematical analysis. It was an understandable—even proper—way to push economics forward. And it achieved a great deal. Its central construct, general equilibrium theory, is not just mathematically elegant; in modeling the

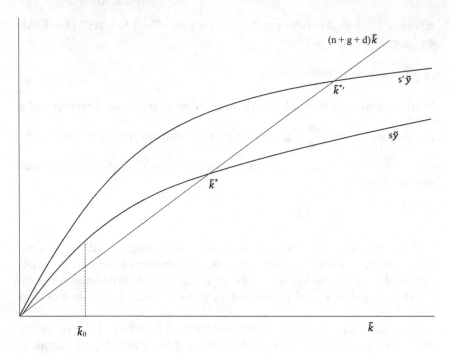

Figure 6.3 Comparative statics of a change in the savings rate from s to s'

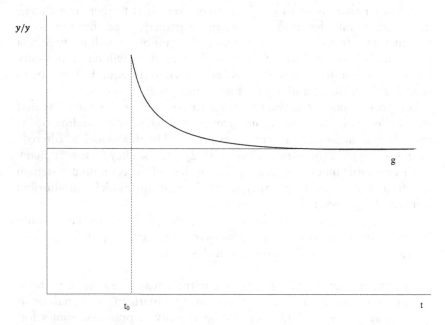

Figure 6.4 A temporary change in the growth rate on the path to the equilibrium growth rate g

economy it recomposes it in our minds, gives us a way to picture it, a way to comprehend the economy in its wholeness. This is extremely valuable, and the same can be said for other equilibrium modelings: of the theory of the firm, of international trade, of financial markets.

But there has been a price for this equilibrium finesse. Economists have objected to it—to the neoclassical construction it has brought about—on the grounds that it posits an idealized, rationalized world that distorts reality, one whose underlying assumptions are often chosen for analytical convenience. I share these objections. Like many economists I admire the beauty of the neoclassical economy; but for me the construct is too pure, too brittle—too bled of reality. It lives in a Platonic world of order, stasis, knowableness, and perfection. Absent from it is the ambiguous, the messy, the real.

On a first reading, the concern here (and in the quotations from the introduction) appears to be as follows. A major motivation for using equilibrium models in understanding a target system—in economics or elsewhere—is that these models provide tractable but also representative summaries of these systems. However, this motivation will be undercut if it turns out that many of the target systems in economics are systems that are not generally in equilibrium (keeping in mind the minimal representationalist view of models assumed here).[16] For then, analyzing the equilibrium states of these models is not representative of the behavior of the target systems—and thus not useful for developing a better understanding these target systems. These systems are never at or near these stable states, so there appears to be little interest in studying these equilibria in any kind of detail.[17]

However, as thus presented, this criticism can be met with an immediate rejoinder: equilibrium systems—whether implicitly or explicitly—can also set out the *dynamics* of the relevant systems, including when they are *not* in equilibrium. This was explicitly noted earlier: equilibrium models describe the states that economic systems will *move towards*. This is important, as it can provide much understanding on its own, without it needing to be the case that these systems actually *reach* their stable states. This is in fact frequently noted by defenders of equilibrium modeling. So, L. Samuelson (2002, pp. 58–59) writes:

> We do not believe that markets are always in equilibrium, just as we do not believe that people are always rational or that firms always maximize profits. But the bulk of our attention is devoted to equilibrium models either because we hope that equilibrium behavior is sufficiently persistent and disequilibrium behavior sufficiently transient that behavior that is robust enough to be an object of study is (approximately) equilibrium behavior, or because studying equilibrium behavior is our best hope for gaining insight into more ephemeral disequilibrium behavior.

The Solow model illustrates this very well. The model describes what will happen if the economy is not in its stable state (e.g. where $\tilde{k}_0 < \tilde{k}^*$): it will approach its stable state at a rate greater than g. Similarly, it describes what happens if the fundamentals of the economy change (e.g. if s changes to s'): the economy shifts from one equilibrium to another in levels of GDP/head, while the growth rate temporarily changes, but then gradually falls back to its equilibrium rate. That is, the model states in what direction the relevant variables will change; whether they adjust gradually or discontinuously; whether they overshoot, undershoot or directly hit their new equilibrium values; etc.

For these reasons, the first reading of the criticism of the use of equilibrium models in economics cannot be considered plausible. The mere fact that many (or even most) economic systems—such as real economies or parts thereof—are rarely directly in an equilibrium does not show that the use of equilibrium models in understanding them needs to be problematic. So, whatever concern there is with the methodology of economics, it cannot just be that real economic systems are always out of their equilibrium.[18]

What other worries with equilibrium modeling in economics exist, though? Two such readings come to mind. First, the concern could be that actual economic systems do not generally *have* stable states (or at least that that there are many time-paths many actual economic systems can take that are not towards any stable states). Depending on exactly where the system happens to be at time t_0, it will end up at very different states at time t_1.[19] Second, the concern could be that, while it is true that at any given moment actual economic systems do move towards a particular stable state, they are subject to frequent shocks that change *which* stable states they are moving towards. If so, then the upshot could be that these systems behave *as if* they had no stable states: they are never on the trajectory towards a given stable state longer than a negligible ("transient" or "ephemeral" in the terminology of the previous quote of Samuelson, 2002) amount. While there are some interesting differences between these two readings, for present purposes, these differences do not matter, and will thus not be further discussed.

Note that for these alternative readings of the worry with equilibrium modeling in economics, this response does not apply. These readings express the thought that economic systems might not be moving *towards* some equilibrium state (or at least not for sufficiently long periods to make it useful to take these stable states into account), not just that they might not be *at* some equilibrium state. This is, at least prima facie, a worry that deserves to be taken seriously. This is especially so, since recent increases in computing power have made dynamic, complex-system-based modeling more feasible than before (Helbing, 2013; Symons & Boschetti, 2012). For this reason, some of the *practical* attractions of the equilibrium-based approach—viz., that it allows for an easy but still representative summary of a complex system like an economy—have been considerably lessened.[20] All in all, it is thus a question that deserves to be taken seriously whether equilibrium modeling in economics is compelling—i.e. whether there is any reason to take seriously the possibility that many or most target systems of economic models are non-equilibrium systems (in one of the previously discussed two senses).

On the face of it, it may appear that determining the answer to this question is a straightforward empirical issue. It would seem that, over time, the success of (representational) equilibrium modeling would show whether (most) economic target systems are equilibrium systems. In line with typical realist views of science (see e.g. Worrall, 1989), it might be thought that, if equilibrium models in economics tend to make novel predictions, many of which are confirmed, and if these models can be used to provide compelling explanations (especially ones that fit well to other kinds explanations) of various economic phenomena, then that would be a reason to think that economic systems are equilibrium systems. If the opposite is true, then that would be a reason to draw the opposite conclusion. Put differently, the plausibility of equilibrium modeling in economics would seem to be something that can be directly read from the empirical success of this way of approaching economic phenomena.

Alas, things are somewhat more complex than that. Interpreting the predictive and explanatory success of equilibrium models need not be straightforward. On the one hand, given the complexities of the kinds of data-generating processes involved in economics, it can be very difficult to determine econometrically whether an equilibrium model or a non-equilibrium model is more successful in accounting for the empirical data (Carlaw & Lipsey, 2012, p. 741). On the other hand, there are some well-known epistemic limitations that economists have to work under (for some classic discussions of these kinds of limitations, see e.g. Scriven, 1956; Machlup, 1961). So, since many of the key economic data concern humans—and often quite intimate facts about humans—there are many ethical and practical limitations on obtaining these data. For example, researchers cannot easily collect data on what specific people earn, exactly what they choose to spend their earnings on, or exactly how specific firms produce their goods.[21] In turn, this lack of data (and related other epistemic difficulties) can go a long way towards explaining away any predictive and explanatory failures in economics. Getting accurate predictions and compelling explanations is difficult *even with the right kinds of models* if one lacks crucial data to which to apply these models. For these reasons, it need not always be obvious whether any empirical failures of equilibrium models in economics are due to the *structure* of these models, rather than to the kinds of limitations economists work under.

Note that it is not the case that it is impossible to resolve these issues with further empirical studies. More data can help clarify to what extent it is indeed the case that it is the lack of helpful data that are the cause of any empirical failures of equilibrium models in economics.[22] Similarly, systematic comparisons between equilibrium and non-equilibrium models can help determine exactly how empirically successful the former are. The point is just that the empirical status of equilibrium modeling in economics is harder to determine than it may at first seem—and that doing so might take significantly more time. In turn, this makes it useful to look for other considerations that may help assess this issue. These other considerations can work together with the empirical assessment of equilibrium modeling in economics to clarify the epistemic status of the latter.

Moreover, one does not have to look too far for these other considerations. Indeed, evolutionary economists have provided theoretical arguments that suggest that we should *expect* that the kinds of complex, interconnected systems modeled in economics will *not* be equilibrium systems.[23] A particularly direct and clear argument of this kind is provided by Galla and Farmer (2013); for some other examples, see e.g. Carlaw and Lipsey (2012) and Arthur (2015).

Galla and Farmer (2013) consider a two-player reiterated game with a large number of moves that are chosen according to a specific type of reinforcement learning. So, at each stage of the game, players pick a move out of a set of N possible moves, with player A's probability of picking a given move M at time t being given by:

$$(3) \quad p(t)^A_M(t) = \frac{e^{\beta Q(t)^A_M}}{\sum_{i=1}^N e^{\beta Q(t)^A_i}},$$

where $Q(t)^A_M$ is a measure of the attractiveness of M to A at t (and exactly similar for player B).[24] That is, moves are chosen according to how attractive they are relative to the other moves the player could choose. Further, $Q(t)^A_M$ (and similarly $Q(t)^B_M$) is updated according to the following rule:

$$(4) \quad Q(t+1)^A_M = (1-\alpha)Q(t)^A_M + \sum_{i=1}^N \Pi^A_{M,i}\, p(t)^B_i$$

where $\Pi^A_{M,i}$ is the payoff to player A of playing move M against player B's move of i. That is, the attractiveness of a move at a given time depends on how attractive that move was in the past, weighted by the extent of the player's memory (α), plus that move's expected current payoff. Galla and Farmer (2013) then do two things: first, they use a normal distribution to randomly assign payoffs to the game, but vary the degree of correlation Γ in the payoffs of the two players. Second, they vary the "memory" of the players—i.e. α—as well as the extent to which the players are drawn to historically successful moves—i.e. β.

The upshot of this is that they find that a large number of the resulting set of games fail to have stable states at all and are characterized by chaotic dynamics. Figure 6.5 (based on figure 6.1 in Galla & Farmer, 2013) makes this clearer:

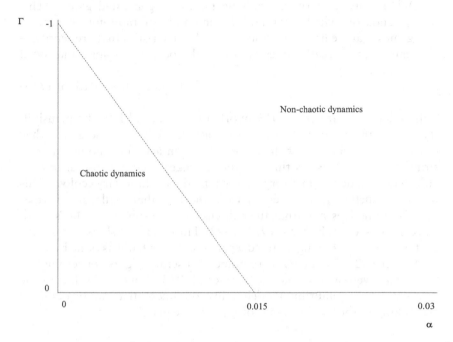

Figure 6.5 Dynamics of Galla and Farmer's (2013) learning games. In these games, β is set at 0.07

What this figure shows is that for competitive games with Γ < 0—which appear to be a key component of economic interactions—even insignificant memory abilities on the parts of the players may lead to chaotic, non-equilibrium dynamics of the game. More than that, Galla and Farmer (2013) note that the situation gets worse as more players are introduced. They write that "our preliminary studies of multiplayer games suggest that these effects become even stronger, and that most parameters lead to chaotic learning dynamics" (Galla & Farmer, 2013, p. 1235). This is important, as many economic interactions—such as financial markets—indeed feature many different "players."

Therefore, Galla and Farmer (2013) suggest that there is good reason to be doubtful about the plausibility of equilibrium modeling in economics. A large number of possible economic interactions seem to be chaotic in nature, and thus, modeling them with equilibrium models is unlikely to be explanatorily or predictively successful. Hence, we here obtain a reason to place the blame on any failures in the empirical success of economic models on the structure of these models themselves.

Is this argument compelling? It would seem that the answer to this question depends on how representative of economic systems Galla and Farmer's (2013) modeling work is. Indeed, they themselves say:

> What can we learn by studying randomly generated games? This depends on whether or not the ensemble of randomly generated games that we have constructed has characteristics that are representative of the "real" games that naturally occur in biology and social science.
>
> (Galla & Farmer, 2013, p. 1233)

Is Galla and Farmer's (2013) work a good inroad into the plausibility of equilibrium modeling in economics, though? Are the games they consider representative of the games played in actual economic interactions? In order to assess this, I want to suggest that it is useful to consider a related debate in a very different field—evolutionary ecology. This approach further gains strength from the fact that Galla and Farmer (2013) themselves note that their discussion is inspired by May's work in ecology (see e.g. R. May, 1974, 1972). However, as I show in the next section, it is useful to dig a little deeper here than what was done by Galla and Farmer (2013): there are a number of useful insights concerning the representativeness of Galla and Farmer's (2013) work—and thus of the plausibility of equilibrium modeling in economics—that can be obtained by looking at the situation in ecology in more detail.

IV. Evolutionary Ecology and the Appeal to Stable Ecosystems

In evolutionary ecology, too, equilibrium models are commonly used, and there, too, this use has become quite controversial. However, it turns out that the discussion in ecology contains several interesting and—so I argue here—telling wrinkles. One way to get at these wrinkles is by considering two key hypotheses in evolutionary ecology:

[1] Actual ecosystems tend to be stable.
[2] More complex and more diverse actual ecosystems are more likely to be stable.

Claim [1] notes that many actual ecosystems—systems of interacting ecological actors and features, such as predators, prey, organisms occupying different niches, and organisms competing for access to the same niches—are equilibrium systems in the sense sketched earlier.[25] In particular, many actual ecosystems appear to show the kind of intermediate stability characterized by the fact that, if the system is (moderately) shocked (e.g. through the disappearance of a major predator), it will tend to return to a state close to its initial, "unshocked" state (Justus, 2008a).[26] Of course, extreme shocks do make the systems unstable, and not all ecosystems are thought to be equilibrium systems in this sense. The point is just that many actual ecosystems, facing realistic shocks, return to their pre-shocked state relatively quickly.

Claim [2] then adds to this the idea that more complex and more diverse actual ecosystems tend to be among the actual ecosystems that are stable in this way (Justus, 2008a). Exactly how *complexity* and *diversity* are to be characterized is a controversial issue (Justus, 2008b, 2008a)—a point to which I return momentarily—but the general idea is that more complex and diverse ecosystems are ones featuring more intricate sets of relationships among a more heterogeneous class of organisms (Justus, 2008a).

Are claims [1] and [2] true? It turns out that, just as in economics, the empirical situation here is somewhat unclear. While both claims initially received some tentative empirical support (Pimentel, 1961; Elton, 1958), this support was obtained by formalizing the concepts of "stability" and "complexity" in ways that were flawed (Justus, 2008b, 2008a). Moreover, it turns out that some very influential theoretical analyses seem to show support for the following two contrasting hypotheses:

[3] Among all the possible ecosystems, a large number are likely to be unstable.

[4] Among all the possible ecosystems, more complex and diverse ones are less likely to be stable than less complex and diverse ones.

The theoretical analyses that underlie [3] and [4] are those by Garnder and Ashby (1970) and especially R. May (1974, 1972) (note also that the latter is exactly the model for the work of Galla & Farmer, 2013, presented in the previous section).[27]

R. May (1972) considered an ecosystem consisting of N types of organisms interacting with each other. Each of these interactions can be positive or negative, and stronger and weaker. This can be represented with a matrix A, each of whose coefficients a_{ij} represents the strength of the interaction between organisms i and j:

$$(5) \quad A = \begin{bmatrix} a_{11} & a_{12} & \cdots & a_{13} \\ a_{21} & a_{22} & \cdots & a_{23} \\ \vdots & \vdots & \ddots & \vdots \\ a_{n1} & a_{n2} & \cdots & a_{nn} \end{bmatrix}$$

R. May (1972) then constrained the diagonal of A to be -1 (to dampen the fluctuations a of given population) and drew values for the non-diagonal

a_{ij} from a distribution with zero mean and variance s^2. He found that this system is almost certainly unstable—in a coarse-grained sense—if

$$(6) \quad s > (NC)^{1/2}$$

where C is the percentage of non-zero a_{ij} in A. This is key, as it suggests that *greater* values of N and C—the measures of diversity and complexity here used—lead to *less* stability. Furthermore, it suggests that a vast number of ecosystems are *unstable*, as very many ecosystems would seem to satisfy (6).

Now, it is important to note that May's work is very controversial (see e.g. Nunney, 1980; Justus, 2008a). For example, some have pointed out that in the set of ecosystems (i.e. the randomly generated A matrices in (5)) surveyed by R. May (1972), there are many that are biologically incoherent for one reason or another (see e.g. Nunney, 1980). Others have noted that the stability concepts used by May are not compelling, and that if more convincing such concepts are used, many more systems—and in particular, more diverse and complex systems—might turn out to be equilibrium systems (see e.g. Justus, 2008b, 2008a).

However, for present purposes, these criticisms do not need to be further assessed. What matters is mostly May's own attitude towards his modeling work. In particular, he notes that:

> Natural ecosystems, whether structurally complex or simple, are the product of a long history of coevolution among their constituent plants and animals. It is at least plausible that such intricate evolutionary processes have, in effect, sought out those relatively tiny and mathematically atypical regions of parameter space which endow the system with long-term stability.
>
> (R. May, 1974, p. 173)

Understanding May's suggestion here is crucial for what follows.

Ecosystems face various kinds of constraints: if the population of a given keystone species (say) is too low or too high, the entire ecosystem will go out of existence. If a significant percentage of the organisms in an ecosystem are—for one reason or another—removed from the ecosystem, that ecosystem may go extinct. More generally, the continued existence of an ecosystem requires the satisfaction of various conditions (including having the right kinds and numbers of member organisms), and this satisfaction cannot be taken for granted. While it is controversial exactly what these conditions are, the fact that there are conditions like these is not (see e.g. Beissinger & Westphal, 1998).

This is a crucial point to note here, as it introduces a kind of sampling bias into the empirical investigation of ecosystems. Over time, most of the ecosystems that remain in existence through the frequent shocks most real systems are subject to—and which can thus be empirically investigated—will tend to be the stable ones. These will be systems that are such that, if shocked, they will return to a viable state (though not necessarily the same one they were in earlier), rather than moving towards a configuration that amounts to their destruction. Equilibrium systems are, by definition, relatively more "bounded" than non-equilibrium systems. Perturbations do not lead these kinds of systems to more or less randomly wander through state space. If there are many regions of state space that are such that, if an ecosystem occupies them, it is no longer viable, then equilibrium systems are therefore in a better position to avoid these regions of state space.[28] Hence, equilibrium ecosystems are less likely to collapse than non-equilibrium ones.

In this way, May shows a way for [1]–[4] to in fact be consistent with each other. While the space of possible ecosystems may contain a large number of unstable ones, the set of *actual* ecosystems is not a representative sample of this space. (Furthermore—though that is not so important here—in the set of actual ecosystems, there can be a correlation between complexity/diversity and stability that is absent in the set of all possible ecosystems.) This is useful, as it is increasingly widely accepted that [1] and [2] *are* plausible within the space of actual ecosystems (though exactly why and how this is to be spelled out is still a matter of debate) (see e.g. Justus, 2008a; Nunney, 1980; Page, 2011). Two further points are important to note about this argument.

First, there is no requirement that the collapse of unstable ecosystems is deterministically guaranteed. To explain the usefulness of equilibrium modeling in ecology, it is enough that there is a significantly strong *positive correlation* between the collapse of an ecosystem and its instability.

Second, as also noted in Chapter 3, it is generally thought that genuine natural selection requires more than just differential survival (Godfrey-Smith, 2009; Brandon, 1990). Hence, this case of ecosystem sorting cannot really be seen as a case of *natural selection*. For this reason, I shall refer to this case as involving ecosystem "sorting" rather than selection. However, this does not affect anything of substance in the argument—and nor does it change its inherently evolutionary nature. What is relevant here is that May's (1974) suggestion provides an evolutionary (though not a selective) reason for why ecosystems are equilibrium systems.

This is key here, as it introduces the possibility that something similar may be true in economics as well. Indeed, Galla and Farmer (2013, p. 1233) hint at this, too:

> [W]hat if it turns out that randomly generated games as we generate them here are not representative of real games? In this case, our approach is still valuable as a null hypothesis that can be used to sharpen understanding of what makes real games special.

The next section spells out this hint in more detail.

V. Lessons From Evolutionary Ecological Modeling for Economic Modeling

Can we learn anything about the plausibility of equilibrium modeling in economics by considering the debate surrounding equilibrium modeling in evolutionary ecology? More specifically, can the linkage between the idea that ecosystems are "sorted" systems and their being equilibrium systems be used as a template with which to approach the discussion in economics? In what follows, I suggest that the answer to this question is a guarded and qualified yes.

There are two reasons for thinking that what makes "real games special" (in Galla and Farmer's terms) is precisely that they are "sorted" systems. In the first place, just like in the case of ecosystems, it is plausible that there are conditions that are such that, if they obtain, economic systems could not persist any longer. For example, once the inflation rate reaches extreme hyperinflationary levels (50% per day, say), economic systems have a hard time persisting (Casella & Feinstein, 1990; F. Taylor, 2013; Capie, 1991). So, economic systems characterized by such inflation rates are likely to sooner or later go out of existence, making the remaining economic systems more likely to be stable.

However, the nature of the conditions required for the persistence of an economic system are even less well understood than those for the persistence of ecosystems. So, it is true that extreme hyperinflation can lead to the collapse of an economy. However, exactly how much inflation is needed for this to occur is not clear, and economic systems can persist for long periods even if they are on an hyperinflationary trajectory (Capie, 1991; Dornbusch & Edwards, 1991). Similarly, it is clear that "bubbles" in various markets can lead to the collapse of these markets (see e.g. the dot-com boom of the late 1990s, the housing bubble in the early 2000s, or the cryptocurrency bubble of the late 2010s). However, there is no clear, widely accepted limit to how high house prices (say) can go. Indeed, it is a widely accepted feature of bubbles and hyperinflationary periods that they are built on the expectations of those in the market, and not on hard exogenous constraints (Sargent & Wallace, 1973).

In short, therefore, it is not clear what the conditions leading to the collapse of economic systems are—and thus, it is not clear *how often* these conditions are obtained. Put differently, it is not clear how large the regions of state space are that make economic systems nonviable, or which regions these are. Accordingly, it is unclear how strong the sorting pressures based on these conditions are. If it is not obvious how easy it is for an economic system to go out of existence, it is not clear how much extinction-based pressure there is on non-equilibrium systems.

However, it is anyway plausible that there is a second set of sorting pressures on economic systems—apart from the viability-focused ones just sketched. In particular, it is reasonable to think that economic systems may be altered *before* they even get to the point at which they are no longer viable. Economic systems are the products of human thought: they are created by humans to harness the fruits of cooperation. For this reason, it is not unreasonable to think that elected politicians or appointed managers are often concerned with shaping them according to their will. An extreme example of this is provided by cases of institutional design, where economic systems (or parts thereof) are designed from scratch (see e.g. Bergemann & Morris, 2005; Hurwicz, 1994; V. L. Smith, 1976). However, various everyday examples exist as well. For example, traders in a market might look for technologies that can improve the ways the relevant market works (mobile phones may be an example of such a technology: Aker & Mbiti, 2010). In general, it is the case that economic systems (unlike ecosystems) are often monitored by people and their features are altered depending on people's perceptions about them.[29]

Furthermore, it is reasonable to think that people are inclined to alter economic systems *in the direction of being equilibrium systems*. That is, it is reasonable to think that people are often concerned precisely with ensuring that economic institutions are structured so that they limit the extent to which the relevant economic system are perturbed after a shock has occurred (as well as with cushioning the systems against shocks to begin with). The reason for this is that, in an important sense, non-equilibrium systems are less predictable than equilibrium systems— precisely because their behavior is less bounded (Werndl, 2009).[30] After a (moderate) shock—or from any point outside their equilibrium state— the behavior of equilibrium systems, by definition, is more bounded than that of non-equilibrium systems. These systems do not wander aimlessly through parameter space, but move towards or stay within (at least) a narrow range of states. In turn, it is plausible that this makes these kinds of systems more user-friendly than non-equilibrium systems. This is so for two reasons.

On the one hand, the future state of an economic system can matter to the decisions of the economic actors in those systems. Economic decisions are often sensitive to where the relevant economic system is headed. For example, the expected value of different investment and savings options can depend on future states of the interest rate or GDP per head. For this reason, equilibrium economic systems can be easier to work with and be a part of than non-equilibrium economic systems: since it is clearer—as more circumscribed—where these systems are going, interacting with and in them is more straightforward and thus attractive.

On the other hand, and for the same sort of reason, non-equilibrium economic systems can be less attractive to economic actors *outside* of those systems than equilibrium systems. For example, to extent that they are less sure about where the relevant economic systems are heading, external actors are less likely to enter into these sorts of economic systems. In turn, to the extent that increasing the number of actors inside the system is desirable—which is often true, as larger systems can provide more opportunities for the economic actors in the system (a classic evolutionary economic point: Schumpeter, 1942; Hayek, 1967)—this provides further reasons to those in a position to change the relevant economic institutions so as to make them more likely to create equilibrium systems.

All of this matters, as it means that there may be an extra set of pressures towards economic systems being equilibrium systems. It is not just that non-equilibrium economic systems may have a harder time persisting in and of themselves (i.e. ending up in regions of state space that make them no longer viable), it is that the non-equilibrium economic systems are more likely to be altered so as to become equilibrium systems to begin with. These kinds of intention-based sorting pressures can thus add to the overall force of the sorting pressures on economic systems to be equilibrium systems.

Now, it needs to be acknowledged that it is controversial exactly which kinds of institutions lead to particularly stable economic systems. However, it also needs to be acknowledged that there is agreement on at least the outlines of these kinds of institutions: for example, well-defined property rights, efficient processes for settling contractual conflicts, and low levels of corruption (Assenova & Regele, 2017; Acemoglu et al., 2001; Reinhart & Rogoff, 2009). This is enough to make it plausible that people can selectively alter or maintain economic systems to make them more stable: enough is known about which institutions lead to more stable economic systems so that the latter can, at least sometimes, be shaped in light of this knowledge—despite the fact that there are still many uncertainties involved.

All in all, this leads to the conclusion that economic systems, too, may face substantial sorting pressures towards being equilibrium systems. In turn, this suggests that modeling these systems with equilibrium models may not be implausible after all. In this way, an evolutionary biological perspective may end up *supporting* equilibrium modeling in economics, not speaking against it, as is commonly supposed (including by Galla & Farmer, 2013; R. Nelson & Winter, 2002; Arthur, 2015; Carlaw & Lipsey, 2012).

At this point, it is important to note that there is no question that the existence of both of these kinds of equilibrium-focused sorting pressures in the economic case can be considered as fully established. This is a key point to which I return momentarily, but for now, it is sufficient to note the claim made so far is just the existence of these kinds of sorting pressures is at least prima facie *plausible*. No claim is made that the existence of these sorting pressures should be seen to be part of the accepted body of economic truths.

Relatedly, it is useful to restate explicitly that the nature of the sorting pressures on ecosystems and economic systems is likely to be fundamentally different. In the ecological case, the pressures towards equilibrium systems are constituted by the literal collapse of non-equilibrium ecosystems. However, as noted previously, this process is only *one* of the sorting processes relevant to the economic case—and probably not the most important one. In economics, another key element of the relevant sorting lies in the intentional agency of the humans interacting with and in these systems. While there is no clear agreement on how different sorting pressures are to be typed—as there is when it comes to whether a giving sorting pressure is a selective pressure—it is reasonable to see the two sorting pressures at stake here as very different from each other. In particular, the intentional and forward-looking sorting of economic systems is very different in nature from the differential extinction of ecosystems.[31] In fact, this point is related to one of the features that is sometimes especially characteristic of the human social sciences: namely, their "reflexivity" (Romanos, 1973; T. May, 1998).[32] Hence, it is reasonable to see these are very different kinds of processes. (Also, as noted in Chapter 1, while this may be overly conservative, this is not greatly problematic; as at worst, we are making a weaker claim than what we could be making.)

This is important, as it implies that the supposed fact that there is differential survival of equilibrium and non-equilibrium systems in ecology does not provide either structural or evidential support for a parallel conclusion in the economic case. It does not provide evidential support, for process/product relations are not what are at stake here.[33] It is not the case that economic systems are the products of ecosystems (at least, the former are not products of the latter in the sense that is relevant here); hence, facts about the former cannot be used to underwrite evidential claims about the latter.

The supposed fact that there is differential survival of equilibrium and non-equilibrium systems in ecology does not provide structural support for a parallel conclusion in the economic case either, since these two cases are not type-identical (unlike, say, in the case of firm evolution studied in Chapter 3). Because of this, there is no entailment relation here between the situation in biology and that in economics. The fact that actual ecosystems are sorted and thus stable does not imply that actual economic systems must also be sorted and thus stable.

In turn, this makes the present application of evolutionary biological thinking in economics different from the sorts of applications that were at the forefront of Chapters 2 through 5. However, this does not mean that the appeal to the discussion in evolutionary ecology is epistemically unimportant for the question of whether economic systems are well approached with equilibrium models. Rather, it just means that this relevance is *heuristic* in nature. As noted in Chapter 1, there are two aspects to this heuristic usefulness (see also Machery, forthcoming; Schulz, 2011b, 2012).

First, the hypothesis that that economic systems are equilibrium systems due to the fact that there are subject to a form of sorting, while far from fully confirmed, is at least *plausible*. These considerations under-write, not the *actual* existence of these kinds of sorting pressures, but their *possible* existence; while the prior arguments are not strong enough to achieve the former, they are strong enough to achieve the latter. A different way of putting this point is that the discussion in evolutionary ecology can be used as the basis for a sort of "plausibility argument" (Worrall, 1989) that suggests that equilibrium modeling in economics *may* be defensible—without though either entailing that the latter is true, nor providing straightforward evidence for it. Rather, it just puts this plausible hypothesis on the table to be further considered.

Second, it is reasonable to see the hypothesis that economic systems are equilibrium systems due to the fact that there are subject to a form of sorting as *novel*. This comes out clearly from the fact that the plausi-bility of equilibrium modeling in economics is in many ways the oppo-site of some of the traditional conclusions in evolutionary economics. As was noted earlier (and also in Chapter 1), much traditional work in evolutionary economics is associated with the *denial* of the usefulness of equilibrium modeling in economics. Hence, the fact that there is evolu-tionary biological *support* for the usefulness of equilibrium modeling in economics underwrites the novelty of the present suggestion (viz., that economic systems are equilibrium systems due to the fact that they are sorted systems).

In short, the appeal to the discussion in evolutionary ecology fulfills the requirements of a useful heuristic device from Chapter 1: it suggests a novel and plausible hypothesis to test further. This, though, does not exhaust the value of the appeal to evolutionary ecology in the discussion of the plausibility of equilibrium modeling in economics. Once it is on the table that economic systems may be equilibrium systems because they are sorted systems, several further novel and at least prima facie plausible hypotheses suggest themselves. Two such hypotheses are especially useful to mention here.

First, older economic systems are more likely to be well approached using equilibrium models than younger ones. The reason for this is that older economic systems—like older ecosystems—have had more time to have been shaped by the sorts of sorting pressures they are subject to. For example, this might be taken to suggest that the economies of the former Soviet Union (which came to their own only in the early 1990s) are less likely to be well approached using equilibrium models than more estab-lished economies like those of France or the United States. (Note again that this should not be taken to mean that older economic systems should be expected to actually be *at* equilibrium. The point is just that the appeal to the discussion in evolutionary ecology suggests that older economic systems are more likely to be *equilibrium systems*.)

Second, the suggestion that economic systems are "sorted" systems also points to the possibility that the better analyzed and understood an economic system is, the more likely it is to be an equilibrium system. In the background of this hypothesis is the idea that the better analyzed and understood an economic system is, the more likely it is that is the kind of system that has been and will be chosen so as to be an equilibrium system. Poorly analyzed and understood economic systems provide fewer affordances to shape them so as to make them convenient for those using them. Conversely, the very fact that an economic system is poorly analyzed and understood also suggests that, in the past, there have not been major attempts to shape it. Since such shaping can proceed piecemeal, it is thus likely (though not guaranteed) that it brings with it an improved understanding of a given system. Two more concrete corollaries of this hypothesis are that (a) richer countries are likely to have more stable economic systems (as the richer a country is, the more likely it is to have the resources to analyze its economic systems—which in turn increases its propensity to become richer), and (b) more heavily centrally managed economic systems (like China's, say) are more likely to be equilibrium systems (as the more centrally managed an economic system is, the more likely it is that steps will be and have been taken to ensure it is an equilibrium system).

Now, similar to what was noted earlier, there is no question that, while these further suggestions have some initial promise, they need to be further investigated before they can be found compelling. So, Reinhart and Rogoff (2009) show that financial crises are common across a wide swathe of countries, but they also show that there are some countries that have managed to both avoid frequent debt defaults and lessen the severity of any banking crises that do occur. Put differently, given what is known about the nature of actual economic systems, the previously suggested hypotheses cannot be brushed aside as non-starters—but neither can they be seen as well-established.

However, this point is not problematic in the present context. Indeed, it brings out the core overarching lesson to be taken away from this chapter: the consideration of discussions in evolutionary biology can be *heuristically* relevant for discussions in economics. The appeal to evolutionary biology can suggest novel and plausible predictions that deserve to be assessed further. This appeal to evolutionary biology can add to the stock of possibilities for making progress in resolving debates in economics (such as the one surrounding the usefulness of equilibrium modeling). While it may not be possible to confirm these hypotheses yet—indeed, these hypotheses need not be true—the value of the appeal to evolutionary biology lies in these hypotheses being considered in the first place.

At this point, a possible objection might come to mind. This objection is based on the concern that the heuristic usefulness of evolutionary biology is just a biographical fact about particular people. It may be that person A finds it useful to consider the discussion surrounding equilibrium modeling in evolutionary ecology to come up with novel hypotheses to test in economics, whereas person B does not find it useful. It is known that there is a lot of variability in the judgements that people make about many different situations (Machery, 2017), so what reason is there to take the present appeal to evolutionary biology as having anything other than biographical interest?

In response, it needs to be noted that there is no question that the appeal to evolutionary ecology is not *necessary* to derive these economic hypotheses (and thus, a fortiori, that this appeal is not necessary for *all* people). However, this is not the issue either. The point here is just that the appeal to the discussion in evolutionary ecology (by someone) is *one way* in which the suggestion that a form of sorting underwrites the plausibility of equilibrium modeling in economics could be derived. That there are also other ways to derive this conclusion does not damage the heuristic value of the appeal to the discussion of equilibrium modeling in evolutionary ecology. As noted in Chapter 1, the observer-dependence of heuristic devices needs to be accepted. One aspect of this observer-dependence is that the same conclusion can be derived using multiple heuristic devices: no one such device is necessarily integral to its derivation. However, this does not imply that the use of one such heuristic device is therefore epistemically irrelevant: it remains the case that, where this particular device was used to suggest novel and plausible hypotheses to test, it is epistemically relevant (though not in a deductive-structural or an inductive-evidential sense).

VI. Conclusions

A major—and classic—worry concerning contemporary economic methodology concerns its overreliance on equilibrium models: i.e. models that analyze an economic system by focusing on the states that it will eventually settle into and to which it will return if disturbed. As I have shown here, though, this worry cannot be understood as the concern that real economic systems are never at equilibrium since equilibrium models can and often do also specify the dynamics of the system as it *approaches* a given stable state, so the fact that the system is not actually at the equilibrium is not greatly worrisome. More worrisome is the fact that the actual economic systems may not have stable states—or, if they have them, that they may be changing so frequently that these states are approached for any significant period of time.

In order to address this worry, I then looked at a somewhat similar discussion in evolutionary ecology. Here, I have shown that an interesting upshot of this discussion is the fact that ecosystems can be seen as "sorted" systems, and that this might account for their being equilibrium systems. I then applied this idea to the economic case to suggest that there may also be sorting pressures towards equilibrium systems in that context. However, these sorting pressures are likely to be at least partially quite different from those in the ecological realm—viz., they are more likely to be driven by intentional and foresighted human intervention. Finally, I have shown that these ideas make several concrete predictions that it is worthwhile to investigate in more detail: (1) older economic systems are more likely to be well approached using equilibrium models than younger ones; (2) richer economic systems are more likely to be well approached using equilibrium models than poorer ones; and (3) more managed economic systems are more likely to be well approached using equilibrium models than less managed ones.

Apart from the substantive interest of this discussion, I have also shown that it has some more general methodological implications. In particular, the comparison of equilibrium modeling in evolutionary ecology and economics makes clearer that the appeal to evolutionary biology in economics can be interesting even if the two cases are neither structurally parallel (as in Chapters 2 and 3) nor evidentially related to each other (as in Chapters 4 and 5). In particular, this appeal can be heuristically useful in the sense that it can suggest useful and novel ideas to explore further. The present discussion surrounding equilibrium modeling in evolutionary ecology and economics brings out one way in which this can work.

Suggested Further Reading

Carlaw and Lipsey (2012) argue that an evolutionary biological, non-equilibrium-based economic perspective is both empirically and theoretically plausible. Galla and Farmer (2013) give an ecologically inspired argument for the fact that many economic interactions are not equilibrium-based. Werndl (2009) is a sophisticated investigation of the link between chaotic systems and unpredictable systems. Odenbaugh (2005) presents heuristic reasons for idealized modeling in ecology. Justus (2008a) is an overview of different measures and concepts of complexity, diversity, and stability used in ecology. Strevens (2003) is an account of the extent to which systems that are chaotic at one level may still be analyzable with non-chaotic models.

Notes

1. This criticism has also been made from the vantage point of physics: see e.g. Buchanan and Vanberg (1991). However, as noted in Chapter 1, the focus is here on connections between economics and evolutionary biology—which, though, is not to say that much of what follows is not also relevant to an "econophysics"-based approach towards these issues.
2. This is also consistent with much that is in Reiss (2012).
3. This is sometimes called "resilience" (Justus, 2008a).
4. The fact that this model involves discrete dynamics is purely for expositional convenience (the same is true for the other exemplary models in the following). As also made clearer in section III, many economic models feature continuous dynamics.

5. Generally, the point at which the basins of attraction meet is an unstable equilibrium. See the next section for more on this. This form of equilibrium is sometimes called "local stability" or "tolerance" (Justus, 2008a).

6. This is sometimes called "resistance" (Justus, 2008a).

7. A classic—if very weak—example of this kind of case is provided by the concept of Lyapunov stability. A state is said to be Lyapunov stable if, for every bounded neighborhood one cares to specify, a neighborhood of perturbations can be found that ensures that the future states of the system will remain within the specified bounded neighborhood. For criticisms of this equilibrium concept, see Justus (2008b, 2008a).

8. A more complex version of this case concerns saddle points: these are equilibria that are stable in one set of dimensions, but unstable in another. However, for present purposes, these need not be further discussed.

9. Note that this does not require that a system is strictly globally stable—only that the system approaches a stable state within that part of its domain that it can actually be expected to be in. I return to this point momentarily.

10. Relatedly, if equilibrium modeling is understood so broadly as to also include perturbation analysis in general (see e.g. Bonnans & Shapiro, 2000), it is not clear there is any other significant modeling approach. Hence, a narrower understand of the kind of equilibrium analysis at stake here is needed. I thank Ulrich Gaehde for useful discussion of this point.

11. While systems that are only partially chaotic may be approachable to some degree using representational equilibrium models, such models cannot *fully* capture them. This is all that is relevant here. Note also that a given abstract/formal system can be an equilibrium or a non-equilibrium system depending on how the relevant parameter values are set: see e.g. R. May (1974) and Benhabib (2008).

12. Note that the source of this being a disequilibrium system does not reside in its being probabilistic in nature. (Assuming a sufficiently lenient time-horizon, the system where q_t undergoes a positive step-change according to this set of probabilistic rules is an equilibrium system: $P(q_{t+1} > q_t) = 0.9$ if $q_t < q_E$, $P(q_{t+1} > q_t) = 0.1$ if $q_t > q_E$, and $P(q_{t+1} > q_t) = P(q_{t+1} < q_t) = 0$ if $q_t = q_E$, with q_E an exogenous parameter.) The switch to probabilistic dynamics was done merely to avoid the impression that the issue at stake here only concern deterministic systems.

13. This thus points to one of the main motivations behind the use of equilibrium models: it is generally (though not always) the case that describing and understanding a system in all of its details is impossible—there are typically too many causal factors underlying a given system to make a full description of the way it behaves impossible. I return to this point briefly in the next section.

14. As will also be made clearer momentarily, the same is true in many parts of evolutionary biology.

15. See also Skyrms (1992).

16. Given the fact economic systems are highly variegated in nature, this criticism is consistent with there being *some* economic systems that are equilibrium systems. The point it tries to make is just that a *significant proportion* of them are not.

17. One possible response to this criticism may be thought to lie in the idea that this point is now widely accepted, and that economists now frequently rely on non-equilibrium models as well (see e.g. Gähde et al., 2013; Helbing, 2013). As the prior remarks suggest, though, it is not obvious to what extent this is true. At any rate, as the rest of this chapter shows, there may well be

ways to defend the reasonableness of equilibrium modeling in economics. For this reason, I here assume that a response to this criticism is still needed.

18. It is of course true that some particular economic models—such as the classic Walrasian ones—lack compelling dynamics (see e.g. Fisher, 1983; Walker, 1987). However, the point is just that this cannot be seen as intrinsic problem of all equilibrium models.

19. If these systems are deterministic, this would imply that they are truly chaotic: Werndl (2009).

20. That said, the heavy reliance on computers—which non-equilibrium modeling comes with—has its own dangers as well (see e.g. Symons & Horner, 2014).

21. Recall that this is also one of the main motivations for appealing to evolutionary biological considerations on the evidential form of evolutionary economics. See Chapter 4.

22. See e.g. Jordà et al. (2017) for an example.

23. It may be thought that Strevens's (2005, 2003) "enion probability analysis" (EPA)—which shows why and how systems displaying chaotic behavior on one level can display non-chaotic behavior at a higher level—also provides an inroad into this issue (though one that favors equilibrium modeling). However, for two reasons, this is not a useful approach to take here. First, it is not entirely clear how EPA-analysis relates to equilibrium modeling specifically. It is entirely possible that a system satisfying its assumptions produces "simple behavior"—i.e. which is describable with few variables (Strevens, 2003, p. 5)—but which is not equilibrium-focused as set out here. Second and more importantly, it is not clear how widely the assumptions of the EPA approach translate into the economic case—a point also noted by Strevens (2003, pp. 351–355). At any rate, the discussion in this paper is entirely consistent with the one in Strevens: as will be made clearer in the following, there may indeed be good reasons to think that economic systems, despite their complexity, are well approached with equilibrium models.

24. In Galla and Farmer's (2013, p. 1232) description of the learning dynamics, they at times appear to switch the labelling for players and moves. The description in the text appears to do justice to what is intended, though.

25. Some ecologists use the term "ecosystem" in a more precise way that contrasts with, say, "community" by including a broader range of elements (including physical ones such as energy): see e.g. Tansley (1935) and McIntosh (1985). However, in this chapter, I use "ecosystem" simply as shorthand for "ecological system": the kinds of systems studied by ecologists—whatever, exactly, these turn out to be.

26. Put differently, these systems have at least "coarse-grained" stable states.

27. The work of R. May (1974, 1972) and Garnder and Ashby (1970) led some ecologists to argue for a paradigm shift in ecology towards non-equilibrium-focused analyses: see e.g. Wiens (1984) and Wu and Loucks (1995).

28. Another way of spelling out this point is in terms of the evolvability or viability of stable ecosystems as opposed to that of non-stable ecosystems. However, since this would introduce a number of further complexities, I will leave this aside here. For more on evolvability, see e.g. Kirschner and Gerhart (1998) and Pigliucci (2008); for more on viability, see e.g. Agmon et al. (2016).

29. Ecosystems can also be created and shaped by humans (see e.g. Mitsch, 2012). However, in the case of ecosystems, this is a recent phenomenon whose extent is unclear. In the case of economic systems, it is an essential feature of these systems (see e.g. Ofek, 2001).

30. This is not to say that equilibrium systems need to be fully or easily predictable. The point is just that they are more predictable than non-equilibrium systems when it comes to their behavior after they are shocked or outside of the equilibrium.
31. The intentional, agency-based sorting pressures can also operate on a much shorter time scale, but it is not clear that this is an essential difference between these two cases.
32. Defenders of niche construction—like Odling-Smee et al. (2003)—might point out that system alteration also happens in the ecological world. However, as was also made clear in Odling-Smee et al. (2003), the intentional, stability-focused system alterations at stake here are different from the ones common in ecology. This is not to say that intentional, stability-focused system alterations could not occur in ecology, only that they are less common there. See Chapter 2 for more on how to understand niche construction.
33. This point is further strengthened by the fact that the existence of equilibrium-focused sorting pressures is controversial in the ecological case to begin with. As noted earlier, R. May (1974, 1972) takes his modeling work to *suggest* that real ecosystems may be a special class of ecosystems sorted out from all the possible ecosystems. However, this suggestion need to be looked at further before it can be considered well-established (see also Briske et al., 2017). In turn, this further underwrites the point that the plausibility of ecosystem sorting cannot be used as evidence for the plausibility of economic system sorting: for the former plausibility is still a matter of contention.

Conclusion

It is now useful to draw together the many different strands of the argument of this book. In the first place, these strands can be grouped into two bundles: a methodological bundle and a substantive bundle.

On the methodological side, I have tried to show that evolutionary economics divides into three distinct but related forms: the structural form, the evidential form, and the heuristic form. While the epistemic strength of the three forms decreases from the first to the last, all three forms can, at least in principle, advance discussions in economics—they just do so differently. The structural form allows for the wholesale translation of findings in evolutionary biology to economics, the evidential form provides reasons for thinking certain economics hypotheses are true, and the heuristic form suggests novel and plausible economic hypotheses to assess further. I have also tried to show that a thorough understanding of how each of the three forms works profits from an understanding of how they differ from each other: the structural form sees economic systems as evolutionary systems, the evidential form sees economic agents as evolved biological entities, and the heuristic form sees some evolutionary biological phenomena and some economic phenomena as having some non-structural features in common. In this way, the structural and evidential forms are distinguished from each other in terms of whether economic actors and systems are seen as evolving systems themselves, or rather as products of evolving systems. The structural and heuristic forms are distinguished from each other in terms of whether the features that the relevant evolutionary biological and economic phenomena have in common are central, individuating ones, or merely peripheral (and possibly observer-dependent) ones. Finally, the evidential and the heuristic forms are distinguished from each other in terms of whether the relationship between the relevant evolutionary biological and economic phenomena is one of cause and effect, or mere (and possibly observer-dependent) similarity.

On the substantive side, I have tried to show the following:

(1) There are circumstances where economic decision-making is structurally parallel to natural selection, but they are restricted to cases where (i) an agent's preferences are sufficiently strongly correlated

with their biological advantageousness, (ii) economic agency is based on preference maximization, (iii) the difference between global and local maximization is not relevant for how natural selection and economic agency work, and (iv) preferences must be able to be seen to have a formal structure that is rich enough to allow for comparisons across preference values *or* only one set of preference/fitness values is appealed to. In turn, this can help clarify human attitudes towards risk in some cases—such as those involving the repeated strategic division of foodstuffs.

(2) There are circumstances where market competition is structurally parallel to natural selection, but these are restricted to cases where (i) firms are able to reproduce, not just grow, (ii) firms differ in features that are inherited by the offspring entities, and (iii) the features that are inherited by offspring entities are relevant for their profitability. In turn, this can help clarify when firms are economic agents of their own: namely, when the market rewards intra-firm cooperation with firm reproduction—modulated by the pressures on the employees of the firm to become cooperative, hard workers.

(3) It is possible to relate data on animal economic decision-making to the debate of whether economic decision-making is constituted by a monistic or a pluralistic decision-making mechanism, and of what the nature of this mechanism is. However, doing so requires a carefully laid out methodological framework and is not straightforward. A look at the currently available data provides some tentative support to the conclusion that economic decision-making is underwritten by a pluralistic set of economic decision-making mechanisms.

(4) An evolutionary biological perspective can partially clarify which differences in the ways in which different groups of humans make economic decisions are *fundamental* and which *superficial*. In particular, evolutionary biological considerations can be used to support the hypothesis that cultural differences in sharing disposition are fundamental in nature. However, evolutionary biological considerations cannot (currently) be used to advance the debate concerning the nature of gender differences in attitudes towards risk.

(5) In evolutionary ecology, it has been suggested that the fact that ecosystems tend to be "sorted" systems might account for their being equilibrium systems. In turn, this idea suggests that something similar may be true in economics: specifically, to the extent that economic systems can also be seen to be sorted systems—e.g. because they have been deliberately designed in a certain way—they are more likely to be equilibrium systems. While it is not yet clear whether this hypothesis is true, it is at least worthy of further consideration.

These two strands of conclusions (the methodological and the substantive) should not be seen to be entirely separate from each other. Rather, they can be brought together to yield two overarching conclusions.

The first major lesson of this book is that appealing to considerations from evolutionary biology really can help advance debates in economics—but that it need not do so either. There are indeed both promises and challenges in bringing together the two subjects. The appeal to considerations from evolutionary biology in advancing debates in economics cannot be brushed aside as a non-starter (as has sometimes been done in related fields, such as evolutionary psychology—see e.g. Richardson, 2007). However, neither can it be seen as a panacea that solves all problems and which makes for the only compelling way forward for economics (as has also been suggested in related fields; Tooby & Cosmides, 1992; see also R. Nelson et al., 2018). The appeal to considerations from evolutionary biology is *one of* the tools contemporary economists have at their disposal—nothing more, but also nothing less.

The second major lesson of the book is that it is possible to say more about what the sources are of the promises and challenges of evolutionary economics. Specifically, the book makes clear when the synthesis of the two subjects is fruitful—and how. So, the book shows that the structural form of evolutionary economics is epistemically strong, but also that it comes with very stringent conditions of applicability. The evidential form is also epistemically compelling—though slightly weaker than the structural form—but it faces the challenge that its inferences rest on somewhat uncertain premises and require significant degrees of internal and external coherence to be plausible. The heuristic form also has the potential to make a useful contribution to the resolution of various economic debates, though its impact is the least epistemically stable and strong of all the three forms.

Where do we go from here? There are two key future directions for this kind of research. First, a useful next step is to spell out the policy implications of the conclusions reached in this book in more detail. As previously noted, several of these conclusions have more or less concrete policy implications. However, when it comes to policy-making, the details are crucial, and getting clearer on these details is thus a natural extension of the conclusions of this book. The second useful step concerns the examination of other cases. As noted in the introduction and in Chapter 1, there are many further examples of the use of evolutionary biological considerations to make progress in economics. Assessing these in more detail—determining their strengths, weaknesses, epistemic value, and policy implications—is thus something that makes for another useful direction of future research.

All in all, I have thus tried to bring out the nature, difficulties, and promises of bringing evolutionary biological considerations to bear on economic debates. More generally, I have tried to bring out something about the nature, difficulties, and promises of interdisciplinary research more generally—and the important role that philosophy plays in this kind of research.

Bibliography

Abdellaoui, M. (2002). A Genuine Rank-Dependent Generalization of the Von Neumann-Morgenstern Expected Utility Theorem. *Econometrica, 70*(2), 717–736.

Abe, H., & Lee, D. (2011). Distributed Coding of Actual and Hypothetical Outcomes in the Orbital and Dorsolateral Prefrontal Cortex. *Neuron, 70,* 731–741.

Acemoglu, D., Johnson, S., & Robinson, J. A. (2001). The Colonial Origins of Comparative Development: An Empirical Investigation. *The American Economic Review, 91*(5), 1369–1401.

Achinstein, P. (2013). *Evidence and Method: Scientific Strategies of Isaac Newton and James Clerk Maxwell.* Oxford: Oxford University Press.

Aghion, P., & Holden, R. (2011). Incomplete Contracts and the Theory of the Firm: What Have We Learned over the Past 25 Years? *Journal of Economic Perspectives, 25,* 181–197.

Agmon, E., Gates, A. J., Churavy, V., & Beer, R. D. (2016). Exploring the Space of Viable Configurations in a Model of Metabolism: Boundary Coconstruction. *Artificial Life, 22,* 153–171.

Aker, J. C., & Mbiti, I. M. (2010). Mobile Phones and Economic Development in Africa. *Journal of Economic Perspectives, 24*(3), 207–232.

Alchian, A. (1950). Uncertainty, Evolution, and Economic Theory. *The Journal of Political Economy, 58,* 211–221.

Alcock, J. (2013). *Animal Behavior: An Evolutionary Approach* (10th ed.). Sunderland: Sinauer Associates.

Aldrich, H. E., Hodgson, G., Hull, D., Knudsen, T., Mokyr, J., & Vanberg, V. J. (2008). In Defence of Generalized Darwinism. *Journal of Evolutionary Economics, 18*(5), 577–596.

Alexander, J. (2007). *The Structural Evolution of Morality.* Cambridge: Cambridge University Press.

Alexander, J. (2009). Evolutionary Game Theory. *The Stanford Encyclopedia of Philosophy* (Summer 2019 Edition), Edward N. Zalta (ed.), https://plato.stanford.edu/archives/sum2019/entries/game-evolutionary/.

Alexander, J. (2012). Decision Theory Meets the Witch of Agnesi. *Journal of Philosophy, 109*(12), 712–727.

Alvard, M. S. (2003). Kinship, Lineage, and an Evolutionary Perspective on Cooperative Hunting Groups in Indonesia. *Human Nature, 14*(2), 129–163.

Angner, E. (2016). *A Course in Behavioral Economics* (2nd ed.). London: Palgrave Macmillan.

Angner, E. (2018). What Preferences Really Are. *Philosophy of Science, 85*(4), 660–681.

Ariew, A., Rice, C., & Rohwer, Y. (2014). Autonomous-Statistical Explanations and Natural Selection. *British Journal for the Philosophy of Science, 66*(3), 635–658.

Arthur, W. B. (2015). *Complexity and the Economy*. Oxford: Oxford University Press.

Ashraf, Q., & Galor, O. (2013). The "Out of Africa" Hypothesis, Human Genetic Diversity, and Comparative Economic Development. *American Economic Review, 103*(1), 1–46.

Assenova, V. A., & Regele, M. (2017). Revisiting the Effect of Colonial Institutions on Comparative Economic Development. *PLoS One, 12*(5), e0177100.

Bajtelsmit, V. L., & VanDerhei, J. L. (1997). Risk Aversion and Pension Investment Choices. In M. S. Gordon (Ed.), *Positioning Pensions for the Twenty-First Century* (pp. 45–66). Philadelphia: University of Pennsylvania Press.

Barberis, N. C. (2013). Thirty Years of Prospect Theory in Economics: A Review and Assessment. *Journal of Economic Perspectives, 27*(1), 173–196.

Barkow, J. (1992). Beneath New Culture Is Old Psychology: Gossip and Social Stratification. In J. Barkow, L. Cosmides, & J. Tooby (Eds.), *The Adapted Mind: Evolutionary Psychology and the Generation of Culture* (pp. 627–637). Oxford: Oxford University Press.

Batterman, R. (2009). Idealization and Modeling. *Synthese, 169*(3), 427–446.

Baum, J., & McKelvey, B. (Eds.). (1999). *Variations in Organizational Science*. London: Sage.

Baumard, N. (in press). Psychological Origins of the Industrial Revolution. *Behavioral and Brain Sciences*, 1–47.

Beall, C. M., Cavalleri, G. L., Deng, L., Elston, R. C., Gao, Y., Knight, J., . . . Zheng, Y. T. (2010). Natural Selection on EPAS1 (HIF2α) Associated with Low Hemoglobin Concentration in Tibetan Highlanders. *Proceedings of the National Academy of Sciences, 107*(25), 11459–11464.

Bechtel, W., & Abrahamsen, A. (2005). Explanation: A Mechanist Alternative. *Studies in the History and Philosophy of the Biological & Biomedical Sciences, 36*, 421–441.

Becker, G. S. (1976). *The Economic Approach to Human Behavior*. Chicago: University of Chicago Press.

Beggan, J. K. (1992). On the Social Nature of Nonsocial Perception: The Mere Ownership Effect. *Journal of Personal and Social Psychology, 62*, 229–237.

Beinhocker, E. D. (2011). Evolution as Computation: Integrating Self-Organization with Generalized Darwinism. *Journal of Institutional Economics, 7*(3), 393–423.

Beissinger, S. R., & Westphal, M. I. (1998). On the Use of Demographic Models of Population Viability in Endangered Species Management. *The Journal of Wildlife Management, 62*(3), 821–841.

Benhabib, J. (2008). Chaotic Dynamics in Economics. In S. Durlauf & L. E. Blume (Eds.), *The New Palgrave Dictionary of Economics* (pp. 745–748). London: Palgrave Macmillan.

Bergemann, D., & Morris, S. (2005). Robust Mechanism Design. *Econometrica, 73*(6), 1771–1813.

Berger, S. (Ed.). (2009). *The Foundations of Non-Equilibrium Economics: The Principle of Circular and Cumulative Causation*. London: Routledge.

Bicchieri, C. (2006). *The Grammar of Society: The Nature and Dynamics of Social Norms*. Cambridge: Cambridge University Press.

Biglan, A., & Cody, C. (2013). Integrating the Human Sciences to Evolve Effective Policies. *Journal of Economic Behavior & Organization, 90S*, S152–S162.

Binmore, K. (1998). *Game Theory and the Social Contract, Vol. II: Just Playing*. Cambridge, MA: MIT Press.

Binmore, K. (2007). *Playing for Real*. Oxford: Oxford University Press.

Binmore, K. (2009). Interpersonal Comparison of Utility. In H. Kincaid (Ed.), *The Oxford Handbook of Philosophy of Economics* (pp. 540–559). Oxford: Oxford University Press.

Birch, J., & Okasha, S. (2014). Kin Selection and Its Critics. *BioScience, 65*(1), 22–32.

Blake, P. R., McAuliffe, K., Corbit, J., Callaghan, T. C., Barry, O., Bowie, A., . . . Warneken, F. (2015). The Ontogeny of Fairness in Seven Societies. *Nature, 528*, 258.

Blanchard, O. J. (1981). Output, the Stock Market, and Interest Rates. *The American Economic Review, 71*(1), 132–143.

Blanchard, O. J. (2007). *Macroeconomics* (7th ed.). Harlow: Pearson.

Bleichrodt, H., & Wakker, P. P. (2015). Regret Theory: A Bold Alternative to the Alternatives. *The Economic Journal, 125*(583), 493–532.

Bonduriansky, R. (2003). Layered Sexual Selection: A Comparative Analysis of Sexual Behaviour within an Assemblage of Piophilid Flies. *Canadian Journal of Zoology, 81*, 479–491.

Bone, J. E., McAuliffe, K., & Raihani, N. J. (2016). Exploring the Motivations for Punishment: Framing and Country-Level Effects. *PLoS One, 11*(8), e0159769.

Bonnans, J. F., & Shapiro, A. (2000). *Perturbation Analysis of Optimization Problems*. New York: Springer.

Booth, A., Cardona-Sosa, L., & Nolen, P. (2014). Gender Differences in Risk Aversion: Do Single-Sex Environments Affect Their Development? *Journal of Economic Behavior & Organization, 99*, 126–154.

Booth, A., & Nolen, P. (2012). Gender Differences in Risk Behaviour: Does Nurture Matter? *The Economic Journal, 122*(558), F56–F78.

Boots, M., & Knell, R. J. (2002). The Evolution of Risky Behaviour in the Presence of a Sexually Transmitted Disease. *Proceedings of the Royal Society of London: Series B: Biological Sciences, 269*(1491), 585–589.

Boudry, M., Vlerick, M., & McKay, R. (2015). Can Evolution Get Us Off the Hook? Evaluating the Ecological Defence of Human Rationality. *Consciousness and Cognition, 33*, 524–535.

Bovens, L. (1992). Sour Grapes and Character Planning. *The Journal of Philosophy, 89*(2), 57–78.

Bovens, L., & Hartmann, S. (2004). *Bayesian Epistemology*. Oxford: Oxford University Press.

Bowles, S., & Gintis, H. (2011). *A Cooperative Species: Human Reciprocity and Its Evolution*. Princeton: Princeton University Press.

Boyd, R., & Richerson, P. (1985). *Culture and the Evolutionary Process*. Chicago: University of Chicago Press.

Boyd, R., & Richerson, P. (2005). *The Origin and Evolution of Cultures*. Oxford: Oxford University Press.

Boyd, R., Richerson, P., & Henrich, J. (2011). The Cultural Niche: Why Social Learning Is Essential for Human Adaptation. *Proceedings of the National Academy of Sciences, 108*(Supplement 2), 10918–10925.

Boyer, P. (2008). Evolutionary Economics of Mental Time Travel? *Trends in Cognitive Science, 12*(6), 219–224.

Boyer, P., & Petersen, M. B. (2017). Folk-Economic Beliefs: An Evolutionary Cognitive Model. *Behavioral and Brain Sciences,* 1–51.

Bradley, R. (2008). Comparing Evaluations. *Proceedings of the Aristotelian Society (Hardback), 108*(1part1), 85–100.

Bradley, R. (2017). *Decision Theory with a Human Face.* Cambridge: Cambridge University Press.

Brandon, R. (1990). *Adaptation and Environment.* Princeton: Princeton University Press.

Branigan, H. P., & Pickering, M. J. (2017). An Experimental Approach to Linguistic Representation. *Behavioral and Brain Sciences, 40,* e282.

Breslin, D. (2011). Reviewing a Generalized Darwinist Approach to Studying Socio-Economic Change. *International Journal of Management Reviews, 13*(2), 218–235.

Briske, D. D., Illius, A. W., & Anderies, J. M. (2017). Nonequilibrium Ecology and Resilience Theory. In D. D. Briske (Ed.), *Rangeland Systems: Processes, Management and Challenges* (pp. 197–227). Cham: Springer International Publishing.

Broder, A. (2000). Assessing the Empirical Validity of the "Take-the-Best" Heuristic as a Model of Human Probabilistic Inference. *Journal of Experimental Psychology: Learning, Memory, and Cognition, 26,* 1332–1346.

Broome, J. (1991a). Utilitarian Metaphysics? In J. Elster & J. E. Roemer (Eds.), *Interpersonal Comparisons of Well-Being* (pp. 70–98). Cambridge: Cambridge University Press.

Broome, J. (1991b). *Weighing Goods.* Oxford: Blackwell.

Brosnan, S. F., Grady, M. F., Lambeth, S. P., Schapiro, S. J., & Beran, M. J. (2008). Chimpanzee Autarky. *PLoS One, 3,* e1518.

Brosnan, S. F., Jones, O. D., Gardner, M., Lambeth, S. P., & Schapiro, S. J. (2012). Evolution and the Expression of Biases: Situational Value Changes the Endowment Effect in Chimpanzees. *Evolution and Human Behavior, 33,* 378–386.

Brosnan, S. F., Jones, O. D., Lambeth, S. P., Mareno, M. C., Richardson, A. S., & Schapiro, S. J. (2007). Endowment Effects in Chimpanzees. *Current Biology, 17,* 1–4.

Brosnan, S. F., Wilson, B. J., & Beran, M. J. (2011). Old World Monkeys Are More Similar to Humans Than New World Monkeys When Playing a Coordination Game. *Proceedings of the Royal Society of London B: Biological Sciences, 279*(1733), https://doi.org/10.1098/rspb.2011.1781.

Bruckner, D. W. (2009). In Defense of Adaptive Preferences. *Philosophical Studies, 142*(3), 307–324.

Bshary, R., & Raihani, N. J. (2017). Helping in Humans and Other Animals: A Fruitful Interdisciplinary Dialogue. *Proceedings of the Royal Society B, 284,* 20170929.

Buchanan, J. M., & Vanberg, V. J. (1991). The Market as a Creative Process. *Economics and Philosophy, 7*(2), 167–186.

Buller, D. (2005). *Adapting Minds.* Cambridge, MA: MIT Press.

Busemeyer, J. R., & Townsend, J. T. (1992). Fundamental Derivations from Decision Field Theory. *Mathematical Social Sciences*, 23, 255–282.

Busemeyer, J. R., & Townsend, J. T. (1993). Decision Field Theory: A Dynamic-Cognitive Approach to Decision Making in an Uncertain Environment. *Psychological Review*, 100(3), 432–459.

Buss, D. M. (2014). *Evolutionary Psychology: The New Science of the Mind* (5th ed.). Boston: Allyn & Bacon.

Buss, D. M., & Hawley, P. H. (Eds.). (2010). *The Evolution of Personality and Individual Differences*. Oxford: Oxford University Press.

Buss, D. M., & Schmitt, D. P. (1993). Sexual Strategies Theory: An Evolutionary Perspective on Human Mating. *Psychological Review*, 100(2), 204–232.

Buss, L. (1987). *The Evolution of Individuality*. Princeton: Princeton University Press.

Bykvist, K. (2010). Can Unstable Preferences Provide a Stable Standard of Well-Being. *Economics and Philosophy*, 26, 1–26.

Byrnes, J. P., Miller, D. C., & Schafer, W. D. (1999). Gender Differences in Risk Taking: A Meta-Analysis. *Psychological Bulletin*, 125(3), 367–383.

Camerer, C. (2007). Neuroeconomics: Using Neuroscience to Make Economic Predictions. *The Economic Journal*, 117, C26–C42.

Campbell, D. T. (1965). Variation, Selection and Retention in Sociocultural Evolution. In H. R. Barringer, G. I. Blanksten, & R. W. Mack (Eds.), *Social Change in Developing Areas: A Reinterpretation of Evolutionary Theory* (pp. 19–49). Cambridge, MA: Schenkman.

Capie, F. H. (Ed.). (1991). *Major Inflations in History*. Cheltenham: Edward Elgar.

Capron, L., & Mitchell, W. (2009). Selection Capability: How Capability Gaps and Internal Social Frictions Affect Internal and External Strategic Renewal. *Organization Science*, 20(2), 294–312.

Carlaw, K. I., & Lipsey, R. G. (2012). Does History Matter?: Empirical Analysis of Evolutionary Versus Stationary Equilibrium Views of the Economy. *Journal of Evolutionary Economics*, 22(4), 735–766.

Carley, K. (1996). A Comparison of Artificial and Human Organizations. *Journal of Economic Behavior and Organization*, 31, 175–191.

Cartwright, N. (1989). *Nature's Capacities and Their Measurement Hardcover*. Oxford: Oxford University Press.

Cartwright, N. (1999). *The Dappled World: A Study of the Boundaries of Science*. Cambridge: Cambridge University Press.

Casella, A., & Feinstein, J. S. (1990). Economic Exchange during Hyperinflation. *Journal of Political Economy*, 98(1), 1–27.

Chakravartty, A. (2017). *Scientific Ontology: Integrating Naturalized Metaphysics and Voluntarist Epistemology*. Oxford: Oxford University Press.

Chao, H.-K., Chen, S.-T., & Millstein, R. L. (Eds.). (2013). *Mechanism and Causality in Biology and Economics*. Dordrecht: Springer.

Chen, M. K., Lakshminarayanan, V., & Santos, L. R. (2006). How Basic Are Behavioral Biases? Evidence from Capuchin Monkey Trading Behavior. *Journal of Political Economy*, 114, 517–537.

Chudek, M., Zhao, W., & Henrich, J. (2013). Culture-Gene Coevolution, Large-Scale Cooperation and the Shaping of Human Social Psychology. In K. Sterelny, R. Joyce, B. Calcott, & B. Fraser (Eds.), *Cooperation and Its Evolution* (pp. 425–458). Cambridge, MA: MIT Press.

Churchland, P. (1985). *Neurophilosophy*. Cambridge, MA: MIT Press.

Clark, A. (1997). *Being There*. Cambridge, MA: MIT Press.

Clarke, E. (2016). A Levels-of-Selection Approach to Evolutionary Individuality. *Biology and Philosophy, 31*, 893–911.

Clutton-Brock, T. (2007). Sexual Selection in Males and Females. *Science, 318*(5858), 1882–1885.

Coase, R. (1937). The Nature of the Firm. *Economica, 4*, 386–405.

Cockburn, A. (2013). Cooperative Breeding in Birds: Toward a Richer Conceptual Framework. In K. Sterelny, R. Joyce, B. Calcott, & B. Fraser (Eds.), *Cooperation and Its Evolution* (pp. 223–246). Cambridge, MA: MIT Press.

Collins, J., Baer, B., & Weber, E. J. (2016). Evolutionary Biology in Economics: A Review. *Economic Record, 92*(297), 291–312.

Contessa, G. (2007). Scientific Representation, Interpretation, and Surrogative Reasoning. *Philosophy of Science, 74*(1), 48–68.

Cooper, W. S. (1987). Decision Theory as a Branch of Evolutionary Theory: A Biological Derivation of the Savage Axioms. *Pscyhological Review, 94*(4), 395–411.

Cooper, W. S. (1989). How Evolutionary Biology Challenges the Classical Theory of Rational Choice. *Biology and Philosophy, 4*, 457–481.

Cooper, W. S. (2001). *The Evolution of Reason: Logic as a Branch of Biology.* Cambridge: Cambridge University Press.

Cosmides, L., & Tooby, J. (1992). Cognitive Adaptations for Social Exchange. In J. Barkow, L. Cosmides, & J. Tooby (Eds.), *The Adapted Mind: Evolutionary Psychology and the Generation of Culture* (pp. 163–228). Oxford: Oxford University Press.

Cronk, L., & Gerkey, D. (2007). Kinship and Descent. In R. Dunbar & L. Barrett (Eds.), *The Oxford Handbook of Evolutionary Psychology* (pp. 463–478). Oxford: Oxford University Press.

Cropp, R., & Gabric, A. (2002). Ecosystem Adaptation: Do Ecosystems Maximize Resilience? *Ecology, 83*(7), 2019–2026.

Cross, I. (2007). Music and Cognitive Evolution. In R. Dunbar & L. Barrett (Eds.), *The Oxford Handbook of Evolutionary Psychology* (pp. 649–667). Oxford: Oxford University Press.

Cubitt, R. P., Munro, A., & Starmer, C. (2004). Testing Explanations of Preference Reversal. *The Economic Journal, 114*(497), 709–726.

Curry, P. (2001). Decision Making under Uncertainty and the Evolution of Interdependent Preferences. *Journal of Economic Theory, 98*, 357–369.

Da Costa, N., & French, S. (2003). *Science and Partial Truth: A Unitary Approach to Models and Scientific Reasoning*. Oxford: Oxford University Press.

Daly, M., & Wilson, M. (1983). *Sex, Evolution and Behavior: Adaptations for Reproduction* (2nd ed.). Boston, MA: Willard Grant Press.

Damuth, J., & Heisler, I. L. (1988). Alternative Formulations of Multi-Level Selection. *Biology and Philosophy, 3*, 407–430.

Darwin, C. (1859). *On the Origin of Species: A Facsimile of the First Edition* (E. Mayr, Ed.). Cambridge, MA: Harvard University Press.

Dawkins, R. (1982). *The Extended Phenotype*. Oxford: Oxford University Press.

Dawkins, R. (1986). *The Blind Watchmaker*. New York: Norton.

Dawkins, R. (1989). *The Selfish Gene* (2nd ed.). Oxford: Oxford University Press.

DeAngelo, G., & Brosnan, S. F. (2013). The Importance of Risk Tolerance and Knowledge When Considering the Evolution of Inequity Responses across the Primates. *Journal of Economic Behavior & Organization, 90S*, S105–S112.

de Finetti, B. (1979). A Short Confirmation of My Standpoint. In M. Allais & O. Hagen (Eds.), *Expected Utility Hypotheses and the Allais Paradox: Contemporary Discussions of the Decisions under Uncertainty with Allais' Rejoinder* (p. 161). Dordrecht: Springer.

Dhami, M. K. (2003). Psychological Models of Professional Decision Making. *Psycholical Science, 14*, 175–180.

Diecidue, E., & Wakker, P. P. (2001). On the Intuition of Rank-Dependent Utility. *The Journal of Risk and Uncertainty, 23*(3), 281–298.

Dopfer, K., & Potts, J. (2004). Evolutionary Realism: A New Ontology for Economics. *Journal of Economic Methodology, 11*(2), 195–212.

Dornbusch, R., & Edwards, S. (1991). The Macroeconomics of Populism. In R. Dornbusch & S. Edwards (Eds.), *The Macroeconomics of Populism in Latin America* (pp. 7–13). Chicago: University of Chicago Press.

Downes, S. (2013). Evolutionary Psychology Is Not the Only Productive Evolutionary Approach to Understanding Consumer Behavior. *Journal of Consumer Psychology, 23*(3), 400–403.

Downes, S. (2016a). Confronting Variation in the Social and Behavioral Sciences. *Philosophy of Science, 83*(5), 909–920.

Downes, S. (2016b). Reform for the Evolutionary Social Sciences or a New Theory of Human Nature? *Metascience, 25*, 479–485.

Downes, S. (2018). Scientific Imperialism and Explanatory Appeals to Evolution in the Social Sciences. In U. Mäki, A. Walsh, & M. Fernández Pinto (Eds.), *Scientific Imperialism: Exploring the Boundaries of Interdisciplinarity* (pp. 224–236). London: Routledge.

Drayton, L. A., Brosnan, S. F., Carrigan, J., & Stoinski, T. S. (2013). Endowment Effect in Gorillas (Gorilla gorilla). *Journal of Comparative Psychology, 127*, 365–369.

Dretske, F. (1981). *Knowledge and the Flow of Information.* Cambridge, MA: MIT Press.

Driscoll, C. (2018). Cultural Evolution and the Social Sciences: A Case of Unification? *Biology and Philosophy, 33*(7).

Dunn, P. O., & Cockburn, A. (1999). Extrapair Mate Choice and Honest Signaling in Cooperatively Breeding Superb Fairy-Wrens. *Evolution, 53*, 938–946.

Eckel, C. C., & Grossman, P. J. (2008). Men, Women and Risk Aversion: Experimental Evidence. In C. Plott & V. Smith (Eds.), *Handbook of Experimental Economics Results* (Vol. 1, pp. 1061–1073). New York: Elsevier.

Eells, E. (1982). *Rational Decision and Causality.* Cambridge: Cambridge University Press.

Ellsberg, D. (1961). Risk, Ambiguity and the Savage Axioms. *Quarterly Journal of Economics, 75*, 643–669.

Elster, J. (1982). The Case for Methodological Individualism. *Theory and Society, 11*, 453–482.

Elster, J. (1983). *Sour Grapes.* Cambridge: Cambridge University Press.

Elton, C. (1958). *The Ecology of Invasions by Animals and Plants.* Chicago: University of Chicago Press.

Enke, S. (1951). On Maximizing Profits: A Distinction between Chamberlin and Robinson. *The American Economic Review, 41*, 566–578.

Epstein, L. G. (1999). A Definition of Uncertainty Aversion. *Review of Economic Studies, 66*, 579–608.

Fama, E. (1980). Agency Problems and the Theory of the Firm. *The Journal of Political Economy, 88,* 288–307.

Fehr, E., & Camerer, C. (2007). Social Neuroeconomics: The Neural Circuitry of Social Preferences. *Trends in Cognitive Science, 11*(10), 419–427.

Fehr, E., & Gaechter, S. (2000). Fairness and Retaliation: The Economics of Reciprocity. *The Journal of Economic Perspectives, 14,* 159–181.

Feintzeig, R. (2016, April 26). Full-Time Hires Buck the Trend at Fast-Food, Retail Chains. *Wall Street Journal.* Retrieved from www.wsj.com/articles/full-time-hires-buck-the-trend-at-fast-food-retail-chains-1461703172

Felsenstein, J. (2004). *Inferring Phylogenies.* Sunderland: Sinauer Associates.

Fessler, D. M. T., Pillsworth, E. G., & Flamson, T. J. (2004). Angry Men and Disgusted Women: An Evolutionary Approach to the Influence of Emotions on Risk Taking. *Organizational Behavior and Human Decision Processes, 95*(1), 107–123.

Fisher, F. M. (1983). *Disequilibrium Foundations of Equilibrium Economics.* Cambridge: Cambridge University Press.

Fiske, A. P. (1992). The Four Elementary Forms of Sociality: Framework for a Unified Theory of Social Relations. *Psychological Review, 99,* 689–723.

Flemming, T. M., Jones, O. D., Mayo, L., Stoinski, T., & Brosnan, S. F. (2012). The Endowment Effect in Orangutans. *International Journal of Comparative Psychology, 25,* 285–298.

Foss, N., & Klein, P. (2008). The Theory of the Firm and Its Critics: A Stocktaking and an Assessment. In E. Brousseau & J.-M. Glachant (Eds.), *Handbook of New Institutional Economics.* Cambridge: Cambridge University Press.

Frank, J. (2003). Natural Selection, Rational Economic Behavior, and Alternative Outcomes of the Evolutionary Process. *The Journal of Socio-Economics, 32*(6), 601–622.

Frank, R. (1988). *Passions within Reason: The Strategic Role of the Emotions.* New York: W. W. Norton.

Frank, R. (2012). *The Darwin Economy: Liberty, Competition, and the Common Good.* Princeton: Princeton University Press.

Friedman, M. (1953). The Methodology of Positive Economics. In *Essays in Positive Economics* (pp. 3–43). Chicago: University of Chicago Press.

Friedman, M., & Savage, L. J. (1948). The Utility Analysis of Choices Involving Risk. *Journal of Political Economy, 56*(4), 279–304.

Frigg, R., & Nguyen, J. (2016). Scientific Representation. *The Stanford Encyclopedia of Philosophy* (Winter 2018 Edition), Edward N. Zalta (ed.), https://plato.stanford.edu/archives/win2018/entries/scientific-representation/.

Futuyma, D. (2009). *Evolution* (2nd ed.). Sunderland, MA: Sinauer Associates.

Gähde, U., Hartmann, S., & Wolf, J. (Eds.). (2013). *Models, Simulations, and the Reduction of Complexity.* Berlin: De Gruyter.

Galanter, J. M., Gignoux, C. R., Oh, S. S., Torgerson, D., Pino-Yanes, M., Thakur, N., . . . Zaitlen, N. (2017). Differential Methylation between Ethnic Sub-Groups Reflects the Effect of Genetic Ancestry and Environmental Exposures. *eLife, 6.*

Gale, J., Binmore, K., & Samuelson, L. (1995). Learning to Be Imperfect: The Ultimatum Game. *Games and Economic Behavior, 8,* 56–90.

Galla, T., & Farmer, J. D. (2013). Complex Dynamics in Learning Complicated Games. *Proceedings of the National. Academy of Sciences, 110*(4), 1232–1236.

Gangestad, S. W., & Simpson, J. A. (2000). The Evolution of Human Mating: Trade-Offs and Strategic Pluralism. *Behavioral & Brain Sciences, 23,* 573–644.

Gardner, A., West, S. A., & Wild, G. (2011). The Genetical Theory of Kin Selection. *Journal of Evolutionary Biology, 24*(5), 1020–1043.

Garnder, M. R., & Ashby, W. R. (1970). Connectance of Large Dynamic (Cybernetic) Systems: Critical Values for Stability. *Nature, 228*, 784.

Gazzaniga, M., Ivry, R. B., & Mangum, G. R. (2009). *Cognitive Neuroscience: The Biology of the Mind* (3rd ed.). New York: W. W. Norton & Company.

Ghiselin, M. (1987). Bioeconomics and the Metaphysics of Selection. *Journal of Social and Biological Structures, 10*, 361–369.

Gibbard, A. (1987). Ordinal Utilitarianism. In E. Feiwel (Ed.), *Arrow and the Foundations of the Theory of Economic Policy* (pp. 135–153). London: Macmillan.

Gibson, R. M. (1996). Female Choice in Sage Grouse: The Roles of Attraction and Active Comparison. *Behavioral Ecology and Sociobiology, 39*, 55–59.

Giere, R. (2004). How Models Are Used to Represent Reality. *Philosophy of Science, 71*(5), 742–752.

Gigerenzer, G. (2007). *Gut Feelings: The Intelligence of the Unconscious*. New York: Penguin.

Gigerenzer, G. (2008). *Rationality for Mortals*. Oxford: Oxford University Press.

Gigerenzer, G., & Selten, R. (Eds.). (2001). *Bounded Rationality: The Adaptive Toolbox*. Cambridge, MA: MIT Press.

Gigerenzer, G., Todd, P. M., & Group, A. R. (2000). *Simple Heuristics That Make Us Smart*. Oxford: Oxford University Press.

Gillespie, J. (1977). Natural Selection for Variances in Offspring Numbers: A New Evolutionary Principle. *The American Naturalist, 111*(981), 1010–1014.

Gillespie, J. (1998). *Population Genetics: A Concise Guide* (2nd ed.). Baltimore: Johns Hopkins University Press.

Gindis, D. (2009). From Fictions and Aggregates to Real Entities in the Theory of the Firm. *Journal of Institutional Economics, 5*, 25–46.

Gintis, H. (2009). *The Bounds of Reason*. Princeton: Princeton University Press.

Glimcher, P. W., Dorris, M. C., & Bayer, H. M. (2005). Physiological Utility Theory and the Neuroeconomics of Choice. *Games and Economic Behavior, 52*(2), 213–256.

Gneezy, U., & Leonard, K. (2009). Gender Differences in Competition: Evidence from a Matrilineal and a Patriarchal Society. *Econometrica, 77*(5), 1637–1664.

Godfrey-Smith, P. (1996). *Complexity and the Function of Mind in Nature*. Cambridge: Cambridge University Press.

Godfrey-Smith, P. (2001). Three Kinds of Adaptationism. In S. H. Orzack & E. Sober (Eds.), *Adaptationism and Optimality* (pp. 335–357). Cambridge: Cambridge University Press.

Godfrey-Smith, P. (2008). Varieties of Population Structure and the Levels of Selection. *British Journal for the Philosophy of Science, 59*, 25–50.

Godfrey-Smith, P. (2009). *Darwinian Populations and Natural Selection*. Oxford: Oxford University Press.

Goldman, A. (1970). *A Theory of Human Action*. Princeton: Princeton University Press.

Goldman, A. (1995). Simulation and Interpersonal Utility. *Ethics, 105*, 709–726.

Goldstein, D. G., & Gigerenzer, G. (2002). Models of Ecological Rationality: The Recognition Heuristic. *Psychological Review, 109*, 75–90.

Goodman, S. N., & Royall, R. (1988). Evidence and Scientific Research. *American Journal of Public Health, 78*(12), 1568–1574.

Goulson, D. (2000). Why Do Pollinators Visit Proportionally Fewer Flowers in Large Patches? *Oikos, 91*, 485–492.

Gowdy, J., Dollimoreb, D. E., Wilson, D. S., & Witt, U. (2013). Economic Cosmology and the Evolutionary Challenge. *Journal of Economic Behavior & Organization, 90S*, S11–S20.

Gowdy, J., Rosser Jr., J. B., & Roy, L. (2013). The Evolution of Hyperbolic Discounting: Implications for Truly Social Valuation of the Future. *Journal of Economic Behavior & Organization, 90S*, S94–S104.

Grafen, A. (1999). Formal Darwinism, the Individual-as-Maximizing-Agent Analogy and Bet-Hedging. *Proceedings of the Royal Society B, 266*, 799–803.

Grafen, A. (2006). Optimization of Inclusive Fitness. *Journal of Theoretical Biology, 238*, 541–563.

Greene, J. (2008). The Secret Joke of Kant's Soul. In W. Sinnott-Armstrong (Ed.), *Moral Psychology* (Vol. 3, pp. 35–79). Cambridge, MA: MIT Press.

Griffin, A. S., & West, S. A. (2002). Kin Selection: Fact and Fiction. *Trends in Ecology and Evolution, 17*, 15–21.

Griffin, J. (1986). *Well Being: Its Meaning, Measurement, and Moral Importance.* Oxford: Clarendon Press.

Griffin, S. R., Smith, M. L., & Seeley, T. D. (2012). Do Honeybees Use the Directional Information in Round Dances to Find Nearby Food Sources? *Animal Behaviour, 83*, 1319–1324.

Griffiths, P. (2005). Review of "Niche Construction". *Biology and Philosophy, 20*, 11–20.

Griffiths, P., & Gray, R. D. (1994). Developmental Systems and Evolutionary Explanation. *The Journal of Philosophy, 91*(6), 277–304.

Griffiths, P., & Stotz, K. (2018). Developmental Systems Theory as a Process Theory. In D. J. Nicholson & J. Dupre (Eds.), *Everything Flows: Towards a Processual Philosophy of Biology* (pp. 225–245). Oxford: Oxford University Press.

Grillner, S., Robertson, B., & Stephenson-Jones, M. (2013). The Evolutionary Origin of the Vertebrate Basal Ganglia and Its Role in Action Selection. *The Journal of Physiology, 591*(22), 5425–5431.

Groenewegen, J., & Vromen, J. (1999). *Institutions and the Evolution of Capitalism: Implications of Evolutionary Economics.* Cheltenham: Edward Elgar.

Gruene-Yanoff, T. (2011). Evolutionary Game Theory, Interpersonal Comparisons and Natural Selection: A Dilemma. *Philosophy and Biology, 26*, 637–654.

Guala, F. (2005). *The Methodology of Experimental Economics.* Cambridge: Cambridge University Press.

Guala, F. (2012). Are Preferences for Real? Choice Theory, Folk Psychology, and the Hard Case for Commonsensible Realism. In A. Lehtinen, J. Kuorikoski, & P. Ylikoski (Eds.), *Economics for Real* (pp. 137–156). London: Routledge.

Gul, F., & Pesendorfer, W. (2008). The Case for Mindless Economics. In A. Caplin & A. Shotter (Eds.), *The Foundations of Positive and Normative Economics* (pp. 3–42). Oxford: Oxford University Press.

Güth, W., & Kliemt, H. (1998). The Indirect Evolutionary Approach: Bridging the Gap between Rationality and Adaptation. *Rationality and Society, 10*(3), 377–399.

Hagen, E., Chater, N., Gallistel, C., Houston, A. I., Kacelnik, A., Kalenscher, T., . . . Stephens, D. (2012). Decision Making: What Can Evolution Do for Us? In

P. Hammerstein & J. R. Stevens (Eds.), *Evolution and the Mechanisms of Decision Making* (pp. 97–126). Cambridge, MA: MIT Press.

Hamilton, W. (1964). The Genetical Theory of Social Behavior. *Journal of Theoretical Biology*, 7, 1–52.

Hammerstein, P., & Hagen, E. H. (2005). The Second Wave of Evolutionary Economics in Biology. *Trends in Ecology and Evolution*, 20(11), 604–609.

Hammerstein, P., & Noe, R. (2016). Biological Trade and Markets. *Philosophical Transactions of the Royal Society B: Biological Sciences*, 371(1687).

Hammerstein, P., & Stevens, J. R. (Eds.). (2012). *Evolution and the Mechanisms of Decision Making*. Cambridge, MA: MIT Press.

Hammond, P. (1991). Interpersonal Comparisons of Utility: Why and How They Are and Should Be Made. In J. Elster & J. E. Roemer (Eds.), *Interpersonal Comparisons of Well-Being* (pp. 200–255). Cambridge: Cambridge University Press.

Hanna, K., Esa, R., Graeme, R., & Per, L. (2002). Sexually Transmitted Disease and the Evolution of Mating Systems. *Evolution*, 56(6), 1091–1100.

Harless, D. W., & Camerer, C. (1994). The Predictive Utility of Generalized Expected Utility Theories. *Econometrica*, 62(6), 1251–1289.

Harsanyi, J. C. (1977). *Rational Behavior and Bargaining Equilibrium in Games and Social Situations*. Cambridge: Cambridge University Press.

Hart, A. G. (2013). Task Partitioning: Is It a Useful Conept. In K. Sterelny, R. Joyce, B. Calcott, & B. Fraser (Eds.), *Cooperation and Its Evolution* (pp. 203–221). Cambridge, MA: MIT Press.

Hart, O. (2008). Reference Points and the Theory of the Firm. *Economica*, 75, 404–411.

Harvey, P. H., & Pagel, M. (1991). *The Comparative Method in Evolutionary Biology*. Oxford: Oxford University Press.

Hausman, D. M. (1989). Arbitrage Arguments. *Erkenntnis*, 30(1/2), 5–22.

Hausman, D. M. (1992). *The Inexact and Separate Science of Economics*. Cambridge: Cambridge University Press.

Hausman, D. M. (1995). The Impossibility of Interpersonal Utility Comparisons. *Mind*, 104, 473–490.

Hausman, D. M. (2008). Why Look Under the Hood? In *The Philosophy of Economics* (pp. 183–187). Cambridge: Cambridge University Press.

Hausman, D. M. (2012). *Preference, Value, Choice, and Welfare*. Cambridge: Cambridge University Press.

Hawks, J., Wang, E. T., Cochran, G. M., Harpending, H. C., & Moyzis, R. K. (2007). Recent Acceleration of Human Adaptive Evolution. *Proceedings of the National Academy of Sciences*, 104(52), 20753–20758.

Hayden, B. Y., Heilbronner, S. R., & Platt, M. L. (2010). Ambiguity Aversion in Rhesus Macaques. *Frontiers in Neuroscience*, 4, 166.

Hayden, B. Y., Pearson, J. M., & Platt, M. L. (2009). Fictive Reward Signals in the Anterior Cingulate Cortex. *Science*, 324, 948–950.

Hayek, F. (1967). *Studies in Philosophy, Politics and Economics*. London: Routledge.

Helbing, D. (2013). Puralistic Modeling of Complex Systems. In U. Gähde, S. Hartmann, & J. Wolf (Eds.), *Models, Simulations, and the Reduction of Complexity* (pp. 53–80). Berlin: De Gruyter.

Helfat, C. E., & Campo-Rembado, M. A. (2016). Intergrative Capabilities, Vertical Integration, and Innovation over Successive Technology Lifecycles. *Organization Science*, 27(2), 249–264.

Heller, Y., & Mohlin, E. (2019). Coevolution of Deception and Preferences: Darwin and Nash Meet Machiavelli. *Games and Economic Behavior, 113*, 223–247.

Henrich, J. (2000). Does Culture Matter in Economic Behavior? Ultimatum Game Bargaining among the Machiguenga of the Peruvian Amazon. *American Economic Review, 90*(4), 973–979.

Henrich, J. (2015). *The Secret of Our Success: How Culture Is Driving Human Evolution, Domesticating Our Species, and Making Us Smarter.* Princeton, NJ: Princeton University Press.

Henrich, J., Boyd, R., Bowles, S., Camerer, C., Fehr, E., Gintis, H., & McElreath, R. (2001). In Search of Homo Economicus: Behavioral Experiments in 15 Small-Scale Societies. *The American Economic Review, 91*(2), 73–78.

Henrich, J., Boyd, R., Bowles, S., Camerer, C., Fehr, E., Gintis, H., . . . Tracer, D. (2005). "Economic Man" in Cross-Cultural Perspective: Behavioral Experiments in 15 Small-Scale Societies. *Behavioral and Brain Sciences, 28*(6), 795–815.

Henrich, J., Heine, S. J., & Norenzayan, A. (2010). The Weirdest People in the World? *Behavioral and Brain Sciences, 33*(2–3), 61–83, discussion 83–135.

Henrich, J., & McElreath, R. (2007). Dual-Inheritance Theory: The Evolution of Human Cultural Capacities and Cultural Evolution. In R. Dunbar & L. Barrett (Eds.), *The Oxford Handbook of Evolutionary Psychology* (pp. 555–570). Oxford: Oxford University Press.

Henrich, J., & McElreath, R. (2011). The Evolution of Cultural Evolution. *Evolutionary Anthropology, 12*, 123–135.

Herrmann-Pillath, C. (2013). *Foundations of Economic Evolution: A Treatise on the Natural Philosophy of Economics.* Cheltenham: Edward Elgar.

Hesse, M. (1963). *Models and Analogies in Science.* London: Sheed & Ward.

Heyes, C. M. (2018). *Cognitive Gadgets: The Cultural Evolution of Thinking.* Cambridge, MA: Harvard University Press.

Hill, B. (2009). Three Analyses of Sour Grapes. In T. Gruene-Yanoff & S. O. Hansson (Eds.), *Preference Change* (pp. 27–56). Dordrecht: Springer.

Hintze, A., & Hertwig, R. (2016). The Evolution of Generosity in the Ultimatum Game. *Scientific Reports, 6*, 34102.

Hirshleifer, J. (1977). Economics from a Biological Viewpoint. *Journal of Law and Economics, 20*, 1–52.

Hitchcock, C., & Sober, E. (2004). Prediction Versus Accommodation and the Risk of Overfitting. *The British Journal for the Philosophy of Science, 55*(1), 1–34.

Hittinger, C. T., & Carroll, S. B. (2007). Gene Duplication and the Adaptive Evolution of a Classic Genetic Switch. *Nature, 449*(7163), 677–681.

Hodgson, G. (1999). *Evolution and Institutions.* Cheltenham: Edward Elgar.

Hodgson, G. (2007a). Meanings of Methodological Individualism. *Journal of Economic Methodology, 14*, 211–226.

Hodgson, G. (2007b). Taxonomizing the Relationship between Biology and Economics: A Very Long Engagement. *Journal of Bioeconomics, 9*(2), 169–185.

Hodgson, G. (2011). A Philosophical Perspective on Contemporary Evolutionary Economics. In J. B. Davis & D. W. Hands (Eds.), *The Elgar Companion to Recent Developments in Economic Methodology* (Vol. 299–318). Cheltenham, UK: Edward Elgar.

206   *Bibliography*

Hodgson, G. (2019). *Evolutionary Economics: Its Nature and Future.* Cambridge: Cambridge University Press.

Hodgson, G., & Knudsen, T. (2010). *Darwin's Conjecture.* Chicago: University of Chicago Press.

Hoover, K. D. (2001). *The Methodology of Empirical Macroeconomics.* Cambridge: Cambridge University Press.

Hoover, K. D. (2009). Microfoundations and the Ontology of Macroeconomics In D. Ross & H. Kincaid (Eds.), *The Oxford Handbook of Philosophy of Economics* (pp. 386–409). Oxford: Oxford University Press.

Hoover, K. D. (2010). Idealizing Reduction: The Microfoundations of Macroeconomics. *Erkenntnis, 73*(3), 329–347.

Hoover, K. D. (2015). Reductionism in Economics: Intentionality and Eschatological Justification in the Microfoundations of Macroeconomics. *Philosophy of Science, 82*(4), 689–711.

Houston, A. I., & McNamara, J. M. (1999). *Models of Adaptive Behaviour: An Approach Based on State.* Cambridge: Cambridge University Press.

Houston, A. I., McNamara, J. N., & Steer, M. (2007). Do We Expect Natural Selection to Produce Rational Behaviour? *Philosophical Transactions of the Royal Society of London B: Biological Sciences, 362,* 1531–1543.

Huerta-Sanchez, E., Jin, X., Asan, Bianba, Z., Peter, B. M., Vinckenbosch, N., . . . Nielsen, R. (2014). Altitude Adaptation in Tibetans Caused by Introgression of Denisovan-Like DNA. *Nature, 512,* 194–197.

Hughes, A. L., & Yeager, M. (1998). Natural Selection at Major Histocompatibility Complex Loci of Vertebrates. *Annual Review of Genetics, 32,* 415–435.

Hull, D. (1988). *Science as a Process: An Evolutionary Account of the Social and Conceptual Development of Science.* Chicago: University of Chicago Press.

Humphrey, N. (1986). *The Inner Eye: Social Intelligence in Evolution.* Oxford: Oxford University Press.

Hurley, S. L. (2005). Social Heuristics That Make Us Smarter. *Philosophical Psychology, 18*(5), 585–612.

Hurwicz, L. (1994). Economic Design, Adjustment Processes, Mechanisms, and Institutions. *Economic Design, 1,* 1–14.

Hutchinson, J. M. C., & Gigerenzer, G. (2005). Simple Heuristics and Rules of Thumb: Where Psychologists and Behavioural Biologists Might Meet. *Behavioural Processes, 69,* 97–124.

Jablonka, E., & Lamb, M. (2005). *Evolution in Four Dimensions: Genetic, Epigenetic, Behavioral, and Symbolic Variation in the History of Life.* Cambridge, MA: MIT Press.

Jackendoff, R. (1999). Possible Stages in the Evolution of the Language Capacity. *Trends in Cognitive Science, 3*(7), 272–279.

Jeffrey, R. (1983). *The Logic of Decision* (2nd ed.). Chicago: University of Chicago Press.

Jianakoplos, N. A., & Bernasek, A. (1998). Are Women More Risk Averse? *Economic Inquiry, 36,* 620–630.

Johnson, D. D. P., Price, M. E., & Van Vugt, M. (2013). Darwin's Invisible Hand: Market Competition, Evolution and the Firm. *Journal of Economic Behavior & Organization, 90S,* S128–S140.

Johnson, J. E. V., & Powell, P. L. (1994). Decision Making, Risk and Gender: Are Managers Different? *British Journal of Management, 5,* 123–138.

Johnson, J. G., & Busemeyer, J. R. (2005). A Dynamic, Stochastic, Computational Model of Preference Reversal Phenomena. *Psychological Review*, *112*(4), 841–861.

Jones, C. (2002). *An Introduction to Economic Growth* (2nd ed.). New York: Norton.

Jordà, O. S., Knoll, K., Kuvshinov, D., Schularick, M., & Taylor, A. M. (2017). The Rate of Return on Everything, 1870–2015. *Federal Reserve Bank of San Francisco Working Paper, 2017–25*, https://doi.org/10.24148/wp2017-25.

Joyce, J. M. (1999). *The Foundations of Causal Decision Theory*. Cambridge: Cambridge University Press.

Joyce, J. M. (2010). A Defense of Imprecise Credences in Inference and Decision Making. *Philosophical Perspectives*, *24*, 281–323.

Justus, J. (2008a). Complexity, Diversity, Stability. In S. Sarkar & A. Plutynski (Eds.), *A Companion to the Philosophy of Biology* (pp. 321–350). Oxford: Blackwell.

Justus, J. (2008b). Ecological and Lyapunov Stability. *Philosophy of Science*, *75*, 421–436.

Kacelnik, A., & Bateson, M. (1996). Risky Theories: The Effects of Variance on Foraging Decisions. *American Zoologist*, *36*, 402–434.

Kagel, J. H., Battalio, R. C., & Green, L. (1995). *Economic Choice Theory: An Experimental Analysis of Animal Behavior* (1st ed.). Cambridge: Cambridge University Press.

Kahneman, D. (2003). Maps of Bounded Rationality: Psychology for Behavioral Economics. *American Economic Review*, *93*, 1449–1475.

Kahneman, D., Knetsch, J., & Thaler, R. H. (1991). Anomalies: The Endowment Effect, Loss Aversion, and Status Quo Bias. *The Journal of Economic Perspectives*, *5*, 193–206.

Kahneman, D., Slovic, P., & Tversky, A. (1982). *Judgment under Uncertainty: Heuristics and Biases*. Cambridge: Cambridge University Press.

Kahneman, D., & Thaler, R. H. (1991). Economic Analysis and the Psychology of Utility: Applications to Compensation Policy. *American Economic Review*, *81*, 341–346.

Kahneman, D., & Tversky, A. (1979). Prospect Theory. *Econometrica*, *47*, 263–291.

Kalenscher, T., Tobler, P., Huijbers, W., Daselaar, S., & Pennartz, C. (2010). Neural Signatures of Intransitive Preferences. *Frontiers in Human Neuroscience*, *4*, 1–14.

Kalenscher, T., & van Wingerden, M. (2011). Why We Should Use Animals to Study Economic Decision Making: A Perspective. *Frontiers in Neuroscience*, *5*, 1–11.

Kanngiesser, P., Santos, L. R., Hood, B. M., & Call, J. (2011). The Limits of Endowment Effects in Great Apes (Pan paniscus, Pan troglodytes, Gorilla gorilla, Pongo pygmaeus). *Journal of Comparative Psychology*, *125*, 436–445.

Karsai, I., & Penzes, Z. (2000). Optimality of Cell Arrangement and Rules of Thumb of Cell Initiation in Polistes Dominulus: A Modeling Approach. *Behavioral Ecology*, *11*, 387–395.

Kenrick, D. T., Griskevicius, V., Sundie, J. M., Li, N. P., Li, Y. J., & Neuberg, S. L. (2009). Deep Rationality: The Evolutionary Economics of Decision Making. *Social Cognition*, *27*(5), 764–785.

Kenrick, D. T., Sadalla, E. K., Groth, G., & Trost, M. R. (1990). Evolution, Traits, and the Stages of Human Courtship: Qualifying the Parental Investment Model. *Journal of Personality, 58*, 97–116.

Kenrick, D. T., & Sundie, J. M. (2005). How Do Cultural Variations Emerge from Universal Mechanisms? *Behavioral and Brain Sciences, 28*(6).

Kenrick, D. T., Sundie, J. M., & Kurzban, R. (2008). Cooperation and Conflict between Kith, Kin, and Strangers: Game Theory by Domains. In C. Crawford & D. Krebs (Eds.), *Foundations of Evolutionary Psychology.* Mahwah, NJ: Erlbaum.

Kirschner, M., & Gerhart, J. (1998). Evolvability. *Proceedings of the National Academy of Sciences, 95*(15), 8420–8427.

Kitcher, P. (1989). Explanatory Unification and the Causal Structure of the World. In P. Kitcher & W. Salmon (Eds.), *Scientific Explanation* (pp. 410–505). Minneapolis: University of Minnesota Press.

Klein, B., Crawford, R., & Alchian, A. (1978). Vertical Integration, Appropriable Rents, and the Competitive Contracting Process. *Journal of Law and Economics, 21*, 297–326.

Klepper, S. (2009). Spinoffs: A Review and Synthesis. *European Management Review, 6*, 159–171.

Knight, F. (1921). *Risk, Uncertainty, and Profit.* Boston, MA: Houghton Mifflin.

Kőszegi, B., & Rabin, M. (2006). A Model of Reference-Dependent Preferences. *Quarterly Journal of Economics, 121*(4), 1133–1165.

Kreps, D. (1990). *A Course in Microeconomic Theory.* Princeton: Princeton University Press.

Kruglanski, A. W., & Gigerenzer, G. (2011). Intuitive and Deliberate Judgments Are Based on Common Principles. *Psychological Review, 118*(1), 97–109.

Kwok, J. J. M., & Lee, D. Y. (2015). Coopetitive Supply Chain Relationship Model: Application to the Smartphone Manufacturing Network. *PLoS One, 10*(7).

Lakshminarayanan, V., Chen, M. K., & Santos, L. R. (2008). Endowment Effect in Capuchin Monkeys. *Philosophical Transactions of the Royal Society of London B: Biological Sciences, 363*, 3837–3844.

Laland, K. N., Odling-Smee, F. J., & Feldman, M. W. (2005). On the Breadth and Significance of Niche Construction. *Biology and Philosophy, 20*, 37–55.

Laland, K. N., & Sterelny, K. (2006). Seven Reasons (Not) to Neglect Niche Construction. *Evolution, 60*, 1751–1762.

Lawson, T. (2006). The Nature of Heterodox Economics. *Cambridge Journal of Economics, 30*, 483–505.

Lee, D., McGreevy, B. P., & Barraclough, D. J. (2005). Learning and Decision Making in Monkeys during a Rock-Paper-Scissors Game. *Cognitive Brain Research, 25*, 416–430.

Lewis, D. (1980). A Subjectivist's Guide to Objective Chance. In R. Jeffrey (Ed.), *Studies in Inductive Logic and Probability* (Vol. 2, pp. 263–293). Berkeley: University of California Press.

Lewontin, R. (1970). The Units of Selection. *Annual Review of Ecology and Systematics*, 1–18.

Lewontin, R. (1982). Organism and Environment. In H. C. Plotkin (Ed.), *Learning, Development and Culture.* New York, NY: Wiley.

Lewontin, R. (1983). Gene, Organism, and Environment. In D. S. Bendall (Ed.), *Evolution from Molecules to Men* (pp. 273–286). Cambridge: Cambridge University Press.

Lewontin, R. (2000). *The Triple Helix: Gene, Organism, and Environment.* Cambridge: Cambridge University Press.

Li, N. P., & Kenrick, D. T. (2006). Sex Similarities and Differences in Preferences for Short-Term Mates: What, Whether, and Why. *Journal of Personality and Social Psychology, 90,* 468–489.

Li, Y. J., Haws, K., & Griskevicius, V. (in press). Parenting Motivation and Consumer Decision Making. *Journal of Consumer Research.*

Li, Y. J., Kenrick, D. T., Griskevicius, V., & Neuberg, S. L. (2012). Economic Decision Biases and Fundamental Motivations: How Mating and Self-Protection Alter Loss Aversion. *Journal of Personality and Social Psychology, 102*(3), 550–561.

List, C., & Pettit, P. (2006). Group Agency and Supervenience. *Southern Journal of Philosophy, 44,* 85–105.

List, C., & Pettit, P. (2011). *Group Agency: The Possibility, Design, and Status of Corporate Agents.* Oxford: Oxford University Press.

Lo, A. W. (2017). *Adaptive Markets.* Princeton, NJ: Princeton University Press.

Loomes, G., Starmer, C., & Sugden, R. (1991). Observing Violations of Transitivity by Experimental Methods. *Econometrica, 59,* 425–439.

Loomes, G., & Sugden, R. (1982). Regret Theory: An Alternative Theory of Choice under Uncertainty. *The Economic Journal, 92*(368), 805–824.

Loomes, G., & Sugden, R. (1987). Some Implications of a More General Form of Regret Theory. *Journal of Economic Theory, 41*(2), 270–287.

Luce, R. D., & Raiffa, H. (1957). *Games and Decisions: Introduction and Critical Survey.* New York: John Wiley.

Lyons, B., Heap, S. H., Hollis, M., Sugden, R., & Weale, A. (1992). *The Theory of Choice: A Critical Guide.* Oxford: Blackwell.

MacDonald, D. H., Kagel, J. H., & Battalio, R. C. (1991). Animals' Choices over Uncertain Outcomes, Further Experimental Results. *Economic Journal, 101,* 1065–1084.

Macfarlan, S. J., & Quinlan, R. J. (2008). Kinship, Family, and Gender Effects in the Ultimatum Game. *Human Nature, 19*(3), 294–309.

Machamer, P., Darden, L., & Craver, C. (2000). Thinking about Mechanisms. *Philosophy of Science, 67*(1), 1–25.

Machery, E. (2009). *Doing without Concepts.* Oxford: Oxford University Press.

Machery, E. (2017). *Philosophy within Its Proper Bounds.* Oxford: Oxford University Press.

Machery, E. (forthcoming). Discovery and Confirmation in Evolutionary Psychology. In J. Prinz (Ed.), *Oxford Handbook of Philosophy of Psychology.* Oxford: Oxford University Press.

Machery, E., & Barrett, H. C. (2006). Debunking Adapting Minds. *Philosophy of Science, 73,* 232–246.

Machlup, F. (1961). Are the Social Sciences Really Inferior? *Southern Economic Journal, 27*(3), 173–184.

Maki, U. (2009). Missing the World: Models as Isolations and Credible Surrogate Systems. *Erkenntnis, 70,* 29–43.

Mäki, U. (2013). Contested Modeling: The Case of Economics. In U. Gähde, S. Hartmann, & J. H. Wolf (Eds.), *Models, Simulations, and the Reduction of Complexity* (pp. 87–106). New York: Walter de Gruyter.

Markman, A., Blok, S., Dennis, J., Goldwater, M., Kim, K., Laux, J., . . . Taylor, E. (2005). Culture and Individual Differences. *Behavioral and Brain Sciences, 28*, 831–831.

Marr, D. (1982). *Vision.* Cambridge, MA: MIT Press.

Marsh, B., & Kacelnik, A. (2002). Framing Effects and Risky Decisions in Starlings. *Proceedings of the National Academy of Sciences, 99*, 3352–3355.

Mas-Colell, A., Whinston, M. D., & Green, J. R. (1996). *Microeconomic Theory.* Oxford: Oxford University Press.

Massimi, M. (2018). Four Kinds of Perspectival Truth. *Philosophy and Phenomenological Research, 96*(2), 342–359.

May, R. (1972). Will a Large Complex System Be Stable? *Nature, 238*(5364), 413–414.

May, R. (1974). *Stability and Complexity in Model Ecosystems.* Princeton, NJ: Princeton University Press.

May, T. (1998). Reflexivity in the Age of Reconstructive Social Science. *International Journal of Social Research Methodology, 1*(1), 7–24.

Maynard Smith, J. (1982). *Evolution and the Theory of Games.* Cambridge: Cambridge University Press.

Maynard Smith, J. (1987). How to Model Evolution. In J. Dupre (Ed.), *The Latest on the Best: Essays on Evolution and Optimality* (pp. 119–131). Cambridge, MA: MIT Press.

Maynard Smith, J., & Szethmary, E. (1995). *The Major Transitions in Evolution.* Oxford: Oxford University Press.

Mayo, D. (1996). *Error and the Growth of Experimental Knowledge.* Chicago: University of Chicago Press.

Mays, H. L., & Hill, G. E. (2004). Choosing Mates: Good Genes Versus Genes That Are a Good Fit. *Trends in Ecology and Evolution, 19*(10), 554–559.

McClamrock, R. (1991). Marr's Three Levels: A Re-Evaluation. *Minds and Machines, 1*(2), 185–196.

McDermott, R., Fowler, J. H., & Smirnov, O. (2008). On the Evolutionary Origin of Prospect Theory Preferences. *The Journal of Politics, 70*(2), 335–350.

McIntosh, R. P. (1985). *The Background of Ecology: Concept and Theory.* Cambridge: Cambridge University Press.

McKerlie, D. (2007). Rational Choice, Changes in Values over Time, and Well-Being. *Utilitas, 19*, 51–72.

McNamara, J. M. (1995). Implicit Frequency-Dependence and Kin Selection in Fluctuating Environments. *Evolutionary Ecology, 9*, 185–203.

Michod, R. (1999). *Darwinian Dynamics: Evolutionary Transitions in Fitness and Individuality.* Princeton: Princeton University Press.

Mill, J. S. (1874). *Essays on Some Unsettled Questions of Political Economy* (2nd ed.). London: Longmans, Green, Reader, and Dyer.

Miller, G. F. (2001). *The Mating Mind: How Sexual Choice Shaped the Evolution of Human Nature.* New York, NY: Anchor Press.

Millikan, R. L. (1984). *Language, Thought, and Other Biological Categories.* Cambridge, MA: MIT Press.

Millikan, R. L. (1989). Biosemantics. *Journal of Philosophy, 86,* 281–297.

Millstein, R. L. (2013). Natural Selection and Causal Productivity. In H.-K. Chao, S.-T. Chen, & R. L. Millstein (Eds.), *Mechanism and Causality in Biology and Economics* (pp. 147–163). Berlin: Springer.

Mithen, S. (2005). *The Singing Neanderthals: The Origins of Music, Language, Mind and Body.* London: Weidenfeld & Nicholson.

Mitsch, W. J. (2012). What Is Ecological Engineering? *Ecological Engineering, 45,* 5–12.

Montoya, E. R., Terburg, D., Bos, P. A., & Van Honk, J. (2012). Testosterone, Cortisol, and Serotonin as Key Regulators of Social Aggression: A Review and Theoretical Perspective. *Motivation and Emotion, 36*(1), 65–73.

Morgan, M. (2012). *The World in the Model.* Cambridge: Cambridge University Press.

Morgan, M., & Morrison, M. (1999). Models as Mediating Instruments. In *Models as Mediators* (pp. 10–37). Cambridge: Cambridge University Press.

Morgenstern, O. (1979). Some Reflections on Utility. In M. Allais & O. Hagen (Eds.), *Expected Utility Hypotheses and the Allais Paradox: Contemporary Discussions of the Decisions under Uncertainty with Allais' Rejoinder* (pp. 175–183). Dordrecht: Springer.

Morillo, C. (1990). The Reward Event and Motivation. *Journal of Philosophy, 87,* 169–186.

Morrison, M. (2015). *Reconstructing Reality: Models, Mathematics and Simulations.* Oxford: Oxford University Press.

Nelson, J. A. (2015). Are Women Really More Risk-Averse Than Men? A Re-Analysis of the Literature Using Expanded Methods. *Journal of Economic Surveys, 29*(3), 566–585.

Nelson, R., Dosi, G., Helfat, C. E., Pyka, A., Saviotti, P. P., Lee, K., . . . Winter, S. (Eds.). (2018). *Modern Evolutionary Economics: An Overview.* Cambridge: Cambridge University Press.

Nelson, R., & Winter, S. (1982). *An Evolutionary Theory of Economic Change.* Cambridge, MA: Belknap Press.

Nelson, R., & Winter, S. (2002). Evolutionary Theorizing in Economics. *Journal of Economic Perspectives, 16*(2), 23–46.

Nettle, D. (2007). Individual Differences. In R. Dunbar & L. Barrett (Eds.), *The Oxford Handbook of Evolutionary Psychology* (pp. 479–490). Oxford: Oxford University Press.

Nichols, S. (2004). *Sentimental Rules: On the Natural Foundations of Moral Judgment.* Oxford: Oxford University Press.

Nickerson, J., & Zenger, T. (2004). A Knowledge-Based Theory of the Firm: The Problem-Solving Perspective. *Organization Science, 15,* 617–632.

Nisbett, R. E., Peng, K., Choi, I., & Norenzayan, A. (2001). Culture and Systems of Thought. *Psychological Review, 108*(2), 291–310.

Nunney, L. (1980). The Stability of Complex Model Ecosystems. *The American Naturalist, 115*(5), 639–649.

Nussbaum, M. C. (2000). *Women and Human Development.* Cambridge: Cambridge University Press.

Nussbaum, M. C. (2001). Symposium on Amartya Sen's Philosophy: 5 Adaptive Preferences and Women's Options. *Economics and Philosophy, 17,* 67–88.

Odenbaugh, J. (2005). Idealized, Inaccurate But Successful: A Pragmatic Approach to Evaluating Models in Theoretical Ecology. *Biology and Philosophy, 20,* 231–255.

Odling-Smee, F. J. (1988). Niche Constructing Phenotypes. In H. C. Plotkin (Ed.), *The Role of Behavior in Evolution* (pp. 73–132). Cambridge, MA: MIT Press.

Odling-Smee, F. J., Laland, K. N., & Feldman, M. W. (2003). *Niche Construction: The Neglected Process in Evolution.* Princeton: Princeton University Press.

Ofek, H. (2001). *Second Nature: Economic Origins of Human Evolution.* Cambridge: Cambridge University Press.

Ofek, H. (2013). MHC-Mediated Benefits of Trade: A Biomolecular Approach to Cooperation in the Marketplace. In K. Sterelny, R. Joyce, B. Calcott, & B. Fraser (Eds.), *Cooperation and Its Evolution* (pp. 175–194). Cambridge, MA: MIT Press.

Okasha, S. (2005). On Niche Construction and Extended Evolutionary Theory. *Biology and Philosophy, 20,* 1–10.

Okasha, S. (2006). *Evolution and the Levels of Selection.* Oxford: Oxford University Press.

Okasha, S. (2007). Rational Choice, Risk Aversion, and Evolution. *The Journal of Philosophy, 104*(5), 217–235.

Okasha, S. (2011). Optimal Choice in the Face of Risk: Decision Theory Meets Evolution. *Philosophy of Science, 78,* 83–104.

Okasha, S., & Binmore, K. (2012). *Evolution and Rationality: Decisions, Cooperation and Strategic Behaviour.* Cambridge: Cambridge University Press.

Orr, H. A. (2007). Absolute Fitness, Relative Fitness, and Utility. *Evolution, 61*(12), 2997–3000.

Orr, H. A. (2009). Fitness and Its Role in Evolutionary Genetics. *Nature Reviews Genetics, 10*(8), 531–539.

Orzack, S. H., & Sober, E. (1994). Optimality Models and the Test of Adaptationism. *The American Naturalist, 143*(3), 361–380.

Ostrom, E. (1990). *Governing the Commons: The Evolution of Institutions for Collective Action.* Cambridge: Cambridge University Press.

Oyama, S. (2000). *Evolution's Eye.* Durgham, NC: Duke University Press.

Page, S. E. (2011). *Diversity and Complexity.* Princeton: Princeton University Press.

Papineau, D. (1987). *Reality and Representation.* Oxford: Blackwell.

Pence, C., & Ramsey, G. (2013a). A New Foundation for the Propensity Definition of Fitness. *British Journal for the Philosophy of Science, 64,* 851–881.

Pence, C., & Ramsey, G. (2013b). A New Foundation for the Propensity Interpretation of Fitness. *British Journal for the Philosophy of Science, 64,* 851–881.

Penke, L. (2010). Bridging the Gap between Modern Evolutionary Psychology and the Study of Individual Differences. In D. M. Buss & P. H. Hawley (Eds.), *The Evolution of Personality and Individual Differences* (pp. 243–279). Oxford: Oxford University Press.

Penrose, E. (1952). Biological Analogies in the Theory of the Firm. *The American Economic Review, 42*(5), 804–819.

Penrose, E. (1959). *The Theory of the Growth of the Firm.* Oxford: Blackwell.

Perini, L. (2010). Scientific Representation and the Semiotics of Pictures. In P. D. Magnus & J. Busch (Eds.), *New Waves in the Philosophy of Science* (pp. 131–154). New York: Macmilan.

Pettit, P. (2003). Groups with Minds of Their Own. In F. Schmitt (Ed.), *Socializing Metaphysics: The Nature of Social Reality* (pp. 167–193). Lanham: Rowman & Littlefield.

Phattanasri, P., Hillel, J. C., & Beer, R. D. (2006). The Dynamics of Associative Learning in Evolved Model Circuits. *Adaptive Behavior, 15*(4), 377–396.

Piccinini, G., & Schulz, A. (2019). The Ways of Altruism. *Evolutionary Psychological Science, 5*, 58–70.

Pigliucci, M. (2008). Is Evolvability Evolvable? *Nature Reviews: Genetics, 9*, 75–82.

Pillsworth, E. G., & Haselton, M. G. (2006). Women's Sexual Strategies: The Evolution of Long-Term Bonds and Extrapair Sex. *Annual Review of Sex Research, 17*(1), 59–100.

Pimentel, D. (1961). Species Diversity and Insect Population Outbreaks. *Annals of the Entomological Society of America, 54*(1), 76–86.

Pinker, S. (1997). *How the Mind Works*. New York: Norton.

Platt, M. L., & Glimcher, P. W. (1999). Neural Correlates of Decision Variables in Parietal Cortex. *Nature, 400*, 233–238.

Polonioli, A. (2015). Stanovich's Arguments against the "Adaptive Rationality" Project: An Assessment. *Studies in History and Philosophy of Science Part C: Studies in History and Philosophy of Biological and Biomedical Sciences, 49*, 55–62.

Potochnik, A. (2010). Explanatory Independence and Epistemic Interdependence: A Case Study of the Optimality Approach. *The British Journal for the Philosophy of Science, 61*(1), 213–233.

Potochnik, A. (2017). *Idealization and the Aims of Science*. Chicago: University of Chicago Press.

Powell, M., & Ansic, D. (1997). Gender Differences in Risk Behaviour in Financial Decision-Making: An Experimental Analysis. *Journal of Economic Psychology, 18*, 605–628.

Pradeu, T. (2012). *The Limits of the Self: Immunology and Biological Identity* (E. Vitanza, Trans.). Oxford: Oxford University Press.

Prugnolle, F., Manica, A., Charpentier, M., Guégan, J. F., Guernier, V., & Balloux, F. (2005). Pathogen-Driven Selection and Worldwide HLA Class I Diversity. *Current Biology, 15*(11), 1022–1027.

Quiggin, J. (1993). *Generalized Expected Utility Theory: The Rank-Dependent Model*. Berlin: Springer.

Rabin, M. (1998). Psychology and Economics. *Journal of Economic Literature, 36*(1), 11–46.

Rabin, M., & Thaler, R. H. (2001). Anomalies: Risk Aversion. *Journal of Economic Perspectives, 15*(1), 219–232.

Radner, R. (2006). Neo-Schumpeterian and Other Theories of the Firm: A Comment and Personal Retrospective. *Industrial and Corporate Change, 15*, 373–380.

Reinhart, C. M., & Rogoff, K. S. (2009). *This Time Is Different: Eight Centuries of Financial Folly*. Princeton: Princeton University Press.

Reiss, J. (2012). Idealization and the Aims of Economics: Three Cheers for Instrumentalism. *Economics and Philosophy, 28*(3), 363–383.

Reiss, J. (2013). *The Philosophy of Economics: A Contemporary Introduction*. London: Routledge.

Resnik, M. (1987). *Choices: An Introduction to Decision Theory*. Minneapolis: University of Minnesota Press.

Richardson, R. (2007). *Evolutionary Psychology as Maladapted Psychology*. Cambridge, MA: MIT Press.

Richmond, P., Mimkes, J., & Hutzler, S. (2013). *Econophysics and Physical Economics*. Oxford: Oxford University Press.

Rickard, M. (1995). Sour Grapes, Rational Desires and Objective Consequentialism. *Philosophical Studies, 80*, 279–303.

Robbins, L. (1932). *An Essay on the Nature and Significance of Economic Science*. London: Palgrave MacMillan.

Robson, A. J. (1996). A Biological Basis for Expected and Non-Expected Utility Theory. *Journal of Economic Theory, 68*, 397–424.

Robson, A. J. (2001a). The Biological Basis of Economic Behavior. *The Journal of Economic Perspectives, 39*(1), 11–33.

Robson, A. J. (2001b). Why Would Nature Give Individuals Utility Functions? *The Journal of Political Economy, 109*(4), 900–914.

Robson, A. J. (2002). Evolution and Human Nature. *The Journal of Economic Perspectives, 16*(2), 89–106.

Robson, A. J. (2003). Evolution and Human Nature: Response from Arthur Robson. *The Journal of Economic Perspectives, 17*(2), 209.

Robson, A. J., & Samuelson, L. (2008). The Evolutionary Foundations of Preferences. In J. Benhabib, A. Bisin, & M. Jackson (Eds.), *The Social Economics Handbook* (pp. 221–310). Amsterdam: Elsevier Press.

Romanos, G. D. (1973). Reflexive Predictions. *Philosophy of Science, 40*(1), 97–109.

Romer, P. M. (1990). Endogenous Technological Change. *The Journal of Political Economy, 98*(5, Part 2: The Problem of Development: A Conference of the Institute for the Study of Free Enterprise Systems), S71–S102.

Rosati, A. G., & Hare, B. (2011). Chimpanzees and Bonobos Distinguish between Risk and Ambiguity. *Biology Letters, 7*, 15–18.

Rosati, A. G., & Hare, B. (2013). Chimpanzees and Bonobos Exhibit Emotional Responses to Decision Outcomes. *PloS One, 8*, e63058.

Rosenberg, A. (1994). *Instrumental Biology*. Chicago: University of Chicago Press.

Rosenberg, A. (2000). The Biological Justification of Ethics: A Best-Case Scenario. In A. Rosenberg (Ed.), *Darwinism in Philosophy, Social Science, and Policy* (pp. 118–136). Cambridge: Cambridge University Press.

Rosenberg, A. (2012). *Philosophy of Social Science* (4th ed.). Boulder, CO: Westview Press.

Ross, D. (2005). *Economic Theory and Cognitive Science: Microfoundations*. Cambridge, MA: MIT Press.

Royall, R. (1997). *Statistical Evidence: A Likelihood Paradigm*. Boca Raton, FL: Chapman and Hall.

Rubin, H. (2018). The Debate over Inclusive Fitness as a Debate over Methodologies. *Philosophy of Science, 85*(1), 1–30.

Russon, A. E., & Andrews, K. (2010). Orangutan Pantomime: Elaborating the Message. *Biology Letters, 7*(4), https://doi.org/10.1098/rsbl.2010.0564.

Saad, G. (2013). Evolutionary Consumption. *Journal of Consumer Psychology, 23*, 351–371.

Saad, G. (in press). On the Method of Evolutionary Psychology and Its Applicability to Consumer Research. *Journal of Marketing Research*.

Samuelson, L. (2001). Introduction to the Evolution of Preferences. *Journal of Economic Theory*, 97(2), 225–230.

Samuelson, L. (2002). Evolution and Game Theory. *The Journal of Economic Perspectives*, 16(2), 47–66.

Samuelson, L., & Swinkels, J. (2006). Information, Evolution, and Utility. *Theoretical Economics*, 1, 119–142.

Samuelson, P. A. (1938). A Note on the Pure Theory of Consumer's Behaviour. *Economica*, 5(17), 61–71.

Santos, L. R., & Chen, K. M. (2009). The Evolution of Rational and Irrational Economic Behavior: Evidence and Insight from a Non-Human Primate Species. In P. W. Glimcher, C. Camerer, E. Fehr, & R. A. Poldrack (Eds.), *Neuroeconomics: Decision Making and the Brain* (1st ed., pp. 81–93). London: Academic Press.

Santos, L. R., & Rosati, A. G. (2015). The Evolutionary Roots of Human Decision Making. *Annual Review of Psychology*, 66(1), 321–347.

Sapienza, P., Zingales, L., & Maestripieri, D. (2009). Gender Differences in Financial Risk Aversion and Career Choices Are Affected by Testosterone. *Proceedings of the National Academy of Sciences*, 106(36), 15268–15273.

Sargent, T. J., & Wallace, N. (1973). Rational Expectations and the Dynamics of Hyperinflation. *International Economic Review*, 14(2), 328–350.

Sarkar, S. (2005). *Molecular Models of Life*. Cambridge, MA: MIT Press.

Satz, D., & Ferejohn, J. (1994). Rational Choice and Social Theory. *The Journal of Philosophy*, 91(2), 71–87.

Savage, L. (1954). *The Foundations of Statistics*. New York: Dover.

Schaffer, J. (2003). Principled Chances. *British Journal for the Philosophy of Science*, 54, 27–41.

Schmalensee, R., & Willig, R. (Eds.). (1989). *Handbook of Industrial Organization*. Amsterdam: North-Holland.

Schmeidler, D. (1989). Subjective Probability and Expected Utility without Additivity. *Econometrica*, 57, 571–587.

Schulz, A. (2008). Risky Business: Evolutionary Theory and Human Attitudes towards Risk: A Reply to Okasha. *Journal of Philosophy*, 105(3), 156–165.

Schulz, A. (2011a). Gigerenzer's Evolutionary Arguments against Rational Choice Theory: An Assessment. *Philosophy of Science*, 78(5), 1272–1282.

Schulz, A. (2011b). Sober & Wilson's Evolutionary Arguments for Psychological Altruism: A Reassessment. *Biology and Philosophy*, 26, 251–260.

Schulz, A. (2012). Heuristic Evolutionary Psychology. In K. S. Plaisance & T. A. C. Reydon (Eds.), *Philosophy of Behavioral Biology* (Vol. 282, pp. 217–234). Dordrecht: Springer.

Schulz, A. (2013). Selection, Drift, and Independent Contrasts: Defending the Methodological Foundations of the FIC. *Biological Theory*, 7(1), 38–47.

Schulz, A. (2015). The Heuristic Defense of Scientific Models: An Incentive-Based Assessment. *Perspectives on Science*, 23, 424–442.

Schulz, A. (2018a). By Genes Alone: A Model Selectionist Argument for Genetical Explanations of Cooperation in Non-Human Organisms. *Biology & Philosophy*, 32, 951–967.

Schulz, A. (2018b). *Efficient Cognition: The Evolution of Representational Decision Making*. Cambridge, MA: MIT Press.

Schumpeter, J. (1942). *Capitalism, Socialism, and Democracy*. New York, NY: Harper & Brothers.

Schumpeter, J. (1959). *The Theory of Economic Development*. Cambridge, MA: Harvard University Press.

Scott-Phillips, T. (2015). Nonhuman Primate Communication, Pragmatics, and the Origins of Language. *Current Anthropology, 56*(1), 56–80.

Scriven, M. (1956). A Possible Distinction between Traditional Scientific Disciplines and the Study of Human Behavior. *Minnesota Studies in the Philosophy of Science, 1*, 330–339.

Seeley, T. D., Visscher, P. K., & Passino, K. M. (2006). Group Decision Making in Honey Bee Swarms. *American Scientist, 94*(3), 220–229.

Sen, A. (1979). Interpersonal Comparisons of Welfare. In M. Boskin (Ed.), *Economics and Human Welfare* (pp. 183–201). New York: Academic Press.

Sen, A. (1985). *Commodities and Capabilities*. Amsterdam: North Holland Publishing.

Sen, A. (1995). Gender Inequality and Theories of Justice. In M. C. Nussbaum & J. Glover (Eds.), *Women, Culture, and Development* (pp. 259–273). Oxford: Clarendon Press.

Sen, A. (1997). Maximization and the Act of Choice. *Econometrica, 65*(4), 745–779.

Shafir, S. (1994). Intransitivity of Preferences in Honey Bees: Support for "Comparative" Evaluation of Foraging Options. *Animal Behaviour, 48*, 55–67.

Shafir, S., Reich, T., Tsur, E., Erev, I., & Lotem, A. (2008). Perceptual Accuracy and Conflicting Effects of Certainty on Risk-Taking Behaviour. *Nature, 453*, 917–920.

Shapiro, L. (2004). *The Mind Incarnate*. Cambridge, MA: MIT Press.

Shettleworth, S. J. (2009). *Cognition, Evolution, and Behavior*. Oxford: Oxford University Press.

Simon, H. A. (1957). *Models of Bounded Rationality*. Cambridge, MA: MIT Press.

Sinn, H.-W. (2003). Weber's Law and the Biological Evolution of Risk Preferences. *Geneva Papers on Risk and Insurance Theory, 28*, 87–100.

Skyrms, B. (1992). Chaos and the Explanatory Significance of Equilibrium: Strange Attractors in Evolutionary Game Dynamics. *PSA: Proceedings of the Biennial Meeting of the Philosophy of Science Association, Vol. 1992, Two: Symposia and Invited Papers*, 374–394.

Skyrms, B. (2004). *The Stag Hunt and the Evolution of Social Structure*. Cambridge: Cambridge University Press.

Skyrms, B. (2010). *Signals: Evolution, Learning, and Information*. Oxford: Oxford University Press.

Smith, V. (2007). *Rationality in Economics: Constructivisit and Ecological Forms*. Cambridge: Cambridge University Press.

Smith, V. L. (1976). Experimental Economics: Induced Value Theory. *The American Economic Review*, 274–279.

Smith, V. L. (2005). Behavioral Economics Research and the Foundations of Economics. *The Journal of Socio-Economics, 34*(2), 135–150.

Smith, V. L. (2008). Economics in the Laboratory. In D. M. Hausman (Ed.), *The Philosophy of Economics* (pp. 334–355). Cambridge: Cambridge University Press.

Sober, E. (1980). Evolution, Population Thinking, and Essentialism. *Philosophy of Science*, 47(3), 350–383.

Sober, E. (1983). Equilibrium Explanation. *Philosophical Studies*, 43, 201–210.

Sober, E. (1984). *The Nature of Selection*. Cambridge: Cambridge University Press.

Sober, E. (1992). Models of Cultural Evolution. In P. Griffiths (Ed.), *Trees of Life* (Vol. 11, pp. 17–39). Dordrecht: Springer.

Sober, E. (1998). Three Differences between Deliberation and Evolution. In P. Danielson (Ed.), *Modeling Rationality, Morality, and Evolution* (pp. 408–422). Oxford: Oxford University Press.

Sober, E. (2000). *Philosophy of Biology* (2nd ed.). Boulder, CO: Westview Press.

Sober, E. (2001). The Two Faces of Fitness. In R. Singh, D. Paul, C. Krimbas, & J. Beatty (Eds.), *Thinking about Evolution: Historical, Philosophical, and Political Perspectives* (pp. 309–321). Cambridge: Cambridge University Press.

Sober, E. (2008). *Evidence and Evolution*. Cambridge: Cambridge University Press.

Sober, E. (2011). *Did Darwin Write the Origin of Species Backwards?* New York: Prometheus Books.

Sober, E. (2014). Evolutionary Theory, Causal Completeness, and Theism: The Case of "Guided" Mutation. In D. Walsh & P. Thompson (Eds.), *Evolutionary Biology: Conceptual, Ethical, and Religious Issues* (pp. 31–44). Cambridge: Cambridge University Press.

Sober, E. (2015). *Ockham's Razors: A User's Manual*. Cambridge: Cambridge University Press.

Sober, E., & Shapiro, L. (2007). Epiphenomenalism: The Dos and the Don'ts. In G. Wolters & P. Machamer (Eds.), *Studies in Causality: Historical and Contemporary* (pp. 235–264). Pittsburgh: University of Pittsburgh Press.

Sober, E., & Wilson, D. S. (1998). *Unto Others: The Evolution and Psychology of Unselfish Behavior*. Cambridge, MA: Harvard University Press.

Sopher, B., & Gigliotti, G. (1993). Intransitive Cycles: Rational Choice or Random Error? *Theory and Decision*, 35, 311–336.

Sprangers, M. A. G., & Schwartz, C. E. (1999). Integrating Response Shift into Health-Related Quality of Life Research: A Theoretical Model. *Social Science & Medicine*, 48, 1507–1515.

Stanley, M. S. (2007). Alfred Marshall and the General Theory of Evolutionary Economics. *History of Economic Ideas*, 15(1), 81–110.

Stanovich, K. E. (2004). *The Robot's Rebellion: Finding Meaning in the Age of Darwin*. Chicago: University of Chicago Press.

Stanovich, K. E., & West, R. F. (2000). Individual Differences in Reasoning: Implications for the Rationality Debate. *Behavioral & Brain Sciences*, 23, 645–665.

Stearns, S. (2000). Daniel Bernoulli (1738): Evolution and Economics under Risk. *Journals of Biosciences*, 25, 221–228.

Steele, C. M. (1988). The Psychology of Self-Affirmation: Sustaining the Integrity of the Self. In L. Berkowitz (Ed.), *Advances in Experimental Social Psychology* (Vol. 22, pp. 261–301). San Diego: Academic.

Sterelny, K. (2003). *Thought in a Hostile World: The Evolution of Human Cognition*. Oxford: Wiley-Blackwell.

Sterelny, K. (2005). Made by Each Other: Organisms and Their Environment. *Biology and Philosophy, 20*, 21–36.

Sterelny, K. (2012a). *The Evolved Apprentice: How Evolution Made Humans Unique*. Cambridge, MA: MIT Press.

Sterelny, K. (2012b). From Fitness to Utility. In K. Binmore & S. Okasha (Eds.), *Evolution and Rationality: Decision, Cooperation, and Strategic Behaviour* (pp. 246–273). Cambridge: Cambridge University Press.

Sterelny, K. (2012c). Language, Gesture, Skill: The Co-Evolutionary Foundations of Language. *Philosophical Transactions of the Royal Society B: Biological Sciences, 367*(1599), 2141.

Sterelny, K. (2016). Cumulative Cultural Evolution and the Origins of Language. *Biological Theory, 11*(3), 173–186.

Sterelny, K., & Griffiths, P. (1999). *Sex and Death*. Chicago: University of Chicago Press.

Stich, S. (2007). Evolution, Altruism and Cognitive Architecture: A Critique of Sober and Wilson's Argument for Psychological Altruism. *Biology and Philosophy, 22*, 267–281.

Stoelhorst, J. W. (2008). The Explanatory Logic and Ontological Commitments of Generalized Darwinism. *Journal of Economic Methodology, 15*(4), 343–363.

Stotz, K. (2006). Molecular Epigenesis: Distributed Specificity as a Break in the Central Dogma. *History and Philosophy of the Life Sciences, 28*, 527–544.

Strevens, M. (2003). *Bigger Than Chaos: Understanding Complexity through Probability*. Cambridge, MA: Harvard University Press.

Strevens, M. (2005). How Are the Sciences of Complex Systems Possible? *Philosophy of Science, 72*(4), 531–556.

Suarez, M. (2004). An Inferential Conception of Scientific Representation. *Philosophy of Science, 71*(5), 767–779.

Sugden, R. (2000). Credible Worlds: The Status of Theoretical Models in Economics. *Journal of Economic Methodology, 7*(1), 1–31.

Sundén, A. E., & Surette, B. J. (1998). Gender Differences in the Allocation of Assets in Retirement Savings Plans. *American Economic Review Papers and Proceedings, 88*, 207–211.

Symons, J., & Boschetti, F. (2012). How Computational Models Predict the Behavior of Complex Systems. *Foundations of Science, 18*(4), 809–821.

Symons, J., & Horner, J. (2014). Software Intensive Science. *Philosophy and Technology, 27*(3), 461–477.

Tan, D. S. (2005). Diversity-Oriented Synthesis: Exploring the Intersections between Chemistry and Biology. *Nature Chemical Biology, 1*, 74.

Tansley, A. G. (1935). The Use and Abuse of Vegetational Concepts and Terms. *Ecology, 16*, 284–307.

Taylor, C. (1971). Interpretation and the Sciences of Man. *The Review of Metaphysics, 25*(1), 3–51.

Taylor, F. (2013). *The Downfall of Money: Germany's Hyperinflation and the Destruction of the Middle Class*. London: Bloomsbury Press.

Thagard, P. (2000). *Coherence in Thought and Action*. Cambridge, MA: MIT Press.

Thaler, R. H. (1980). Toward a Positive Theory of Consumer Choice. *Journal of Economic Behavior & Organization, 1*, 39–60.

Thaler, R. H. (2000). From Homo Economicus to Homo Sapiens. *Journal of Economics Perspectives*, *14*, 133–141.

Thomas, M. G., Ji, T., Wu, J., He, Q., Tao, Y., & Mace, R. (2018). Kinship Underlies Costly Cooperation in Mosuo Villages. *Royal Society Open Science*, *5*, http://doi.org/10.1098/rsos.171535

Thornhill, R., Fincher, C. L., & Aran, D. (2009). Parasites, Democratization, and the Liberalization of Values across Contemporary Countries. *Biological Reviews*, *84*(1), 113–131.

Thrall, P. H., Antonovics, J., & Dobson, A. P. (2000). Sexually Transmitted Diseases in Polygynous Mating Systems: Prevalence and Impact on Reproductive Success. *Proceedings of the Royal Society of London, Series B: Biological Sciences*, *267*(1452), 1555–1563.

Todd, P., & Gigerenzer, G. (2007). Mechanisms of Ecological Rationality: Heuristics and Environments That Make Us Smart. In L. Barrett & R. Dunbar (Eds.), *Oxford Handbook of Evolutionary Psychology* (pp. 197–210). Oxford: Oxford University Press.

Todd, P. M. (2001). Fast and Frugal Heuristics for Environmentally Bounded Minds. In G. Gigerenzer & R. Selten (Eds.), *Bounded Rationality: The Adaptive Toolbox* (pp. 51–70). Cambridge, MA: MIT Press.

Tooby, J., & Cosmides, L. (1992). The Psychological Foundations of Culture. In J. Barkow, L. Cosmides, & J. Tooby (Eds.), *The Adapted Mind* (pp. 19–136). Oxford: Oxford University Press.

Toon, A. (2012). *Models as Make-Belief: Imagination, Fiction, and Scientific Representation*. Houndmills: Palgrave MacMillan.

Trefts, E., Gannon, M., & Wasserman, D. H. (2017). The Liver. *Current Biology*, *27*(21), R1147–R1151.

Trivers, R. (1972). Parental Investment and Sexual Selection. In B. Campbell (Ed.), *Sexual Selection and the Descent of Man: The Darwinian Pivot* (pp. 136–179). Piscataway, NJ: Transaction.

Tsai, R.-C., & Bockenholt, U. (2006). Modeling Intransitive Preferences: A Random-Effects Approach. *Journal of Mathematical Psychology*, *50*, 1–14.

Tversky, A., & Kahneman, D. (1992). Advances in Prospect Theory: Cumulative Representation of Uncertainty. *Journal of Risk and Uncertainty*, *5*(4), 297–323.

van Frassen, B. (2008). *Scientific Representation: Paradoxes of Perspective*. Oxford: Oxford University Press.

Varian, H. R. (1992). *Microeconomic Analysis* (3rd ed.). New York: Norton & Company.

Veblen, T. (1898). Why Is Economics Not an Evolutionary Science? *The Quarterly Journal of Economics*, *12*(4), 373–397.

Voorhoeve, A., Binmore, K., Stefansson, A., & Steward, L. (2016). Ambiguity Attitudes, Framing, and Consistency. *Theory and Decision*, *81*, 313–337.

Vorms, M. (2011). Representing with Imaginary Models: Formats Matter. *Studies in History and Philosophy of Science Part A*, *42*, 287–295.

Vrba, E. (1984). What Is Species Selection? *Systematic Zoology*, *33*, 318–328.

Vromen, J. (2001). The Human Agent in Evolutionary Economics. In J. Laurent & J. Nightingale (Eds.), *Darwinism and Evolutionary Economics* (pp. 184–208). Cheltenham: Edward Elgar.

Vromen, J. (2009). Advancing Evolutionary Explanations in Economics: The Limited Usefulness of Tinbergen's Four-Question Classification. In H. Kincaid

(Ed.), *The Oxford Handbook of Philosophy of Economics* (pp. 337–368). Oxford: Oxford University Press.

Wade Hands, D. (2012). Realism, Commonsensibles, and Economics: The Case of Contemporary Revealed Preference Theory. In A. Lehtinen, J. Kuorikoski, & P. Ylikoski (Eds.), *Economics for Real: Uskali Mäki and the Place of Truth in Economics* (pp. 156–178). London: Routledge.

Wagner, G. P., & Altenberg, L. (1996). Perspective: Complex Adaptations and the Evolution of Evolvability. *Evolution, 50*(3), 967–976.

Waite, T. (2001). Intransitive Preferences in Hoarding Gray Jays (Perisoreus Canadensis). *Behavioral Ecology and Sociobiology, 50*, 116–121.

Wakker, P. P. (2010). *Prospect Theory for Risk and Ambiguity*. Cambridge: Cambridge University Press.

Walker, D. A. (1987). Walras's Theories of Tatonnement. *Journal of Political Economy, 95*(4), 758–774.

Walsh, D. (2015). *Organisms, Agency, and Evolution*. Cambridge: Cambridge University Press.

Watkins, J. (1952). The Principle of Methodological Individualism. *The British Journal for the Philosophy of Science, 3*, 186–189.

Weisberg, M. (2007). Three Kinds of Idealization. *Journal of Philosophy, 104*, 639–659.

Weisberg, M. (2013). *Simulation and Similarity: Using Models to Understand the World*. Oxford: Oxford University Press.

Welsch, H. (2005). Adaptation of Tastes to Constraints. *Theory and Decision, 57*, 379–395.

Werndl, C. (2009). What Are the New Implications of Chaos for Unpredictability? *The British Journal for the Philosophy of Science, 60*, 195–220.

Wessinger, C. A., & Rausher, M. D. (2014). Predictability and Irreversibility of Genetic Changes Underlying Flower Color Change in Penstemon Barbatus. *Evolution, 68*, 1058–1070.

West, S. A., El Mouden, C., & Gardner, A. (2011). Sixteen Common Misconceptions about the Evolution of Cooperation in Humans. *Evolution and Human Behavior, 32*(4), 231–262.

West, S. A., Griffin, A. S., & Gardner, A. (2007). Social Semantics: Altruism, Cooperation, Mutualism, Strong Reciprocity and Group Selection. *Journal of Evolutionary Biology, 20*(2), 415–432.

West, S. A., Griffin, A. S., & Gardner, A. (2008). Social Semantics: How Useful Has Group Selection Been? *Journal of Evolutionary Biology, 21*(1), 374–385.

White, R. (2003). The Epistemic Advantage of Prediction over Accommodation. *Mind, 112*(448), 653–683.

Whiten, A., & Byrne, R. W. (Eds.). (1997). *Machiavellian Intelligence II: Extensions and Evaluations*. Cambridge: Cambridge University Press.

Wiens, J. A. (1984). On Understanding a Nonequilibrium World: Myth and Reality in Community Patterns and Processes. In D. R. Strong, D. Simberloff, L. Abele, & A. B. Thistle (Eds.), *Ecological Communities: Conceptual Issues and the Evidence* (pp. 439–458). Princeton: Princeton University Press.

Williamson, O. (1971). The Vertical Integration of Production: Market Failure Considerations. *American Economic Review, 61*, 112–123.

Williamson, O. (1979). Transaction Cost Economics: The Governance of Contractual Relations. *Journal of Law and Economics, 22*, 233–261.

Wilson, D. S. (2008). Social Semantics: Toward a Genuine Pluralism in the Study of Social Behaviour. *Journal of Evolutionary Biology, 21*(1), 368–373.

Wilson, D. S., Gowdy, J. M., & Rosser, J. B., Jr. (2013). Rethinking Economics from an Evolutionary Perspective. *Journal of Economic Behavior & Organization, 90S*, S1–S2.

Wilson, D. S., Ostrom, E., & Cox, M. E. (2013). Generalizing the Core Design Principles for the Efficacy of Groups. *Journal of Economic Behavior & Organization, 90*, S21–S32.

Wilson, M., & Daly, M. (1985). Competitiveness, Risk-Taking, and Violence: The Young Male Syndrome. *Ethology & Sociobiology, 6*, 59–73.

Wimsatt, W. (2007). *Re-Engineering Philosophy for Limited Beings*. Cambridge, MA: Harvard University Press.

Winter, S. (1988). On Coase, Competence, and the Corporation. *Journal of Law, Economics, and Organization, 4*, 163–180.

Winterhaler, B. (2007). Risk and Decision-Making. In R. Dunbar & L. Barrett (Eds.), *The Oxford Hanbook of Evolutionary Psychology* (pp. 433–445). Oxford: Oxford University Press.

Witt, U. (2003). *The Evolving Economy*. Cheltenham: Edward Elgar.

Witt, U. (2008). What Is Specific about Evolutionary Economics? *Journal of Evolutionary Economics, 18*, 547–575.

Witztum, A. (2012). The Firm, Property Rights, and Methodological Individualism: Some Lessons from J. S. Mill. *Journal of Economic Methodology, 19*, 339–355.

Worrall, J. (1989). Structural Realism: The Best of Both Worlds? *Dialectica, 43*(1/2), 99–124.

Wu, J., & Loucks, O. L. (1995). From Balance of Nature to Hierarchical Patch Dynamics: A Paradigm Shift in Ecology. *Quarterly Review of Biology, 70*, 439–466.

Wylie, A. (1994). Evidential Constraints: Pragmatic Objectivism in Archaeology. In M. Martin & L. McIntyre (Eds.), *Readings in the Philosophy of Social Science* (pp. 747–765). Cambridge, MA: MIT Press.

Wylie, A. (2002). *Thinking from Things: Essays in the Philosophy of Archaeology*. Berkeley, CA: University of California Press.

Young, H. P. (1998). *Individual Strategy and Social Structure: An Evolutionary Theory of Institutions*. Princeton: Princeton University Press.

Zahavi, A. (1977). The Cost of Honesty (Further Remarks on the Handicap Principle). *Journal of Theoretical Biology, 67*, 603–605.

Zeelenberg, M. (1999). Anticipated Regret, Expected Feedback, and Behavioral Decision Making. *Journal of Behavioral Decision Making, 12*, 93–106.

Zimmerman, D. (2003). Sour Grapes, Self-Abnegation, and Character Building. *The Monist, 86*, 220–241.

Index

Printed in the United States
by Baker & Taylor Publisher Services